Microbia

Microbia

A JOURNEY INTO THE UNSEEN WORLD AROUND YOU

Eugenia Bone

RODALE.

Microbia

From the Greek,

mikros = *small,* bios = *life*

RODALE *wellness*

Live happy. Be healthy. Get inspired.

Sign up today to get exclusive access to our authors, exclusive bonuses, and the most authoritative, useful, and cutting-edge information on health, wellness, fitness, and living your life to the fullest.

Visit us online at RodaleWellness.com
Join us at RodaleWellness.com/Join

Rodale books may be purchased for business or promotional use or for special sales. For information, please e-mail: BookMarketing@Rodale.com.

Page 99: "Hymn to Mycology," music by Hugh Aitken and lyrics by W. Multer. Printed by permission of the New York Mycological Society.

Printed in the United States of America

Rodale Inc. makes every effort to use acid-free ∞, recycled paper ♻.

Book design by Amy King

Library of Congress Cataloging-in-Publication Data is on file with the publisher.

ISBN 978-1-62336-735-0

Distributed to the trade by Macmillan

2 4 6 8 10 9 7 5 3 1 hardcover

To my fellow students and professors at E3B,

ancora imparo.

Contents

CONTENTS

Introduction

When I was working on my book about mushrooms, *Mycophilia*, I asked the mycologist Tom Volk whether bacteria lived inside fungi the way they live in us.

"If you ask me," he said, "I think there are probably bacteria in everything."

That quote lingered in my mind for years. If bacteria (and, I would later learn, other microscopic organisms) live in everything, did that mean all living things are connected by microbes?

I started looking for the answer in books about microbiology. That's where I hit my first snag. I had no microbiology background. Actually, I had almost no biology background at all. I was an English major in college not only because I loved to read and write but also because I was convinced I was hopeless at science and math. So the books I read may have contained the answer, but if they did, I couldn't see it.

That's why I went back to college, at 55 years old, to study biology. Going back to school turned out to be a kind of unraveling of my ego, where I had to deal with being bad at something all the time. It wasn't fun, but it turned out to be necessary, because humility is the entry point for understanding nature. One of the many things I learned in college was the deeper you look, the more complex life is.

I also learned that learning is not something you age out of. In fact, it can change everything, no matter how old you are. I went to college to study biology in order to expand my perception of life. And it did, but not in the way I expected. I found out that life itself is a vast conspiracy of microbes.

Microbiology, the study of organisms too small to see with the naked eye, is really difficult to comprehend at 55 or any other age. You can't use your senses to perceive these tiny life-forms. Unless you have a microscope, you have no primary observation of them, only secondary sensing. You can see the burp of methane bubbles in Los Angeles's greasy La Brea Tar Pits, but you can't see the archaea that are producing the gas. It's challenging to describe microscopic organisms with words for the same reason. Throughout the process of writing this book, I kept losing track of my voice, the truth-speak that comes directly from the writer's personality. And it occurred to me my problems were founded in the very limited number of adjectives I could use. I mean, the descriptive tools of my trade—sight, sound, smell, and touch—don't really apply to a bacterium. It seemed like my writerly challenges were synonymous with the challenges I faced in understanding the biology of the microscopic world. My worldview, which includes my way of communicating, is locked into a scale relative to my experience. But the microscopic world operates on a very different scale.

That's a problem, because we can understand nature and ourselves in a deeper way through the lens of a microscope. New discoveries about the impacts of bacteria on our lives occur every day, and while the headlines grab some of us, many of these breakthroughs seem beyond the capacity of most people to understand. But microbiology is where it is happening. This is the age of bacteria, said the evolutionary biologist Stephen Jay Gould, "as it was in the beginning, is now, and ever shall be."

Microbiology is like a foreign country. It is very difficult to get around without a basic vocabulary. That's what the biology course I took at Columbia University (my alma mater: I graduated from Barnard College during the Reagan administration) gave me. I acquired the tools necessary to read most papers by microbiologists and understand their presentations. I only developed tourist biologese—I couldn't learn enough in a year to master weedy, acronym-stuffed papers—but it was enough to get around. And by the second semester, something began to change. I started to see that life, every aspect of life, is sustained by microbes.

I experienced revelation after revelation about the impact of microbes on our lives. For example, microbes link the nonliving and living spheres of the planet. They convert chemicals in the atmosphere into food that can travel up the food chain. And they maintain the balance of chemicals on the planet. If lions and elephants went extinct, life would still go on. But as the biologist Tom Curtis wrote in the journal *Nature Reviews Microbiology*, "If we accidentally poisoned the last two species of ammonia-oxidizing bacteria, that would be another matter."

I learned that inside our cells reside the remnants of ancient bacteria that convert the oxygen we breathe into energy. They are called mitochondria, and without them, we wouldn't exist. I realized that on the microscopic level, the whole idea of species just falls apart because the pace of microbial evolution is so fast that by the time a scientist has identified a bacterial species and picked out a name, its progeny may have evolved into something else.

I found out that mats of microbes, not unlike the orange muck of an intertidal zone, are complex communities composed of microscopic friends and enemies, food makers and degraders of the dead. Ancient microbial mats were the protosoil; they made the colonization of terrestrial Earth possible. I learned that soil is living and dying and dead microbes interacting with a matrix of minerals, upgraded with living and dying and dead plants and animals. Without microbes, soil is just dirt.

Another revelation was the notion that feeding and fighting were invented by microbes, and they have been doing those jobs for every organism that came after them on the timeline of life. Microbes deliver nutrients and provide defenses for fungi, plants, and animals. All complex life evolved with microbes that do these jobs. It's the microbes living on our skin that keep pathogens at bay; it's the microbes in our guts that break down the plants we eat.

Indeed, my whole understanding of how to take care of my body changed. What we eat feeds the microbes in us. Eat a lot of sugar and you feed microbes that ferment sugar, and their population numbers rise. Sugar fermenters produce molecular by-products like lactic acid, which cause tooth decay. Eat a lot of kale and other species thrive, species that produce other by-products, some of which we depend on to make everything from hormones to neurotransmitters. Our food choices affect the population dynamics of our microbes, which in turn affect us.

As my knowledge of microbiology deepened, I became more adept at teasing out the commercial hype and fearmongering in the news. Flesh-eating bacteria, for example, are rare; they're not the microbes to worry about even though an infection by them is truly horrific. The ones to worry about are antibiotic-resistant microbes, which experts estimate will kill 10 million people every year by 2050 unless we find some new medicines. Probiotics, live bacteria and fungi that increase the population of particular species in your gut, work if you suffer from certain kinds of diarrhea, but beyond that, *meh*. And prebiotics are just bacteria (and fungi) food, mainly fiber that feeds those microbes that produce healthful benefits. An apple is a prebiotic.

I came to understand that everything that lives has a microbiome, multiple microbiomes, in fact. There are microbiomes all over plants, microbiomes of the seed and the fruit and the leaves, and microbiomes all over us, in our noses and our ears (one fellow, after dealing with a persistent bacterial infection in one ear, finally had the bright idea to transfer some wax from his healthy ear to his infected ear, and within days his infection disappeared), in our belly buttons, on our hands, in our mouths, between our toes, under our arms, in our genitals—everywhere.

Microbes even define family. Each of us travels through life in a unique cloud of microscopic debris that we shed from our skin and hair and breath and clothes and farts, an aura composed of a million microscopic particles. Depending on how much time and what degree of contact a person has, two or more people's airborne microbial clouds may homogenize. Couples share more microbes than roommates, and families with young children share the most of all.

As I was nearing the end of this book, I met an artist named Andrew Cziraki, who explores social issues using science to construct art forms. I'd heard about a piece he was working on called The Holy Water Project. Part of the project included collecting samples of water from holy water fonts, the basins of blessed water often placed inside the entrance of Catholic churches, into which parishioners dunk their fingers before making the sign of the cross. He conducted DNA analyses of each sample in order to determine what organisms were present. He was on the lookout for something that would illuminate the relationship between the community and the church through an analysis of the water. He found bacteria and yeast, and human DNA, but also DNA that correlated with the locations of the New York City churches: carp DNA in the Chinatown church, parasites like roundworms and tapeworms in the church near Pennsylvania Station, "which kind of made sense because there are a lot of homeless people sleeping there," he said.

Andrew is interested in this because holy water is where people drop off and pick up microbes. The metaphor of the church, that religion binds the community, is literal here. "The community shares itself in this font of microbes," he said. The font is a symbol of sacrament, of purification, "and yet," said Andrew, "what it really does is transmit. I learned it doesn't matter who or what you are. We are all connected."

All along I'd been sensing it: microbes are bridge organisms that connect the living and nonliving, the soil and plants, the plants and people, and people with each other. But now, I understand. We are all connected to everything by microbial life. We hardly perceive the connections, but we are living them, every day.

Microbia tracks my year studying biology—what I learned and what it was like learning it. College life today differs from when I was young. A lot has changed. The science is new, the technology is new, and the pressures and stresses are on a level I don't recall. But much is the same. Students still get morally outraged, they still protest, they still have wickedly funny senses of humor. And everyone is still looking for love, or at least that age-old exchange of microbes we call sex.

When I first registered for classes, I was frustrated to discover I couldn't take courses in microbiology without studying biology and chemistry first. I didn't want to wait a few years before I could dig into what most interested me, so I augmented my biology classes with a pretty vigorous schedule of off-campus reading, interviews, seminars, and lectures. The science I report on comes from all those sources.

But as the mycologist Nicholas Money wrote in his book *The Amoeba in the Room*, "The more we probe, the more we see." I am quite sure that by the time you read this book, there will be many new microbiological discoveries, probably very significant ones. The flood of data is accelerating, and I could have continued to write indefinitely. So obviously, I had to make compromises. For starters, I didn't write about ocean microbiology; nor did I explore the microbiology of different soils, or get into comparative microbiomes between different kinds of animals and plants, or look in depth into the effects of climate change on microbial life, which is significant. For example, researchers have found that the warming climate leads to loss of microbiome diversity in common lizards, which can undermine the lizard's ability to survive environmental conditions such as, well, a warming climate. All these are fascinating and important subjects, and lots of great science is being done.

But I chose to focus on the immediate and familiar: farmland, corn, us.

I had a great deal of assistance on this book, from my professors at Columbia and from many microbiologists in the field who kindly clarified—and then clarified again—what was for me very challenging material. At one point over lunch at a hummus restaurant in New York City, Nicholas Money pointed out that maybe it was something of an advantage to come at microbiology as a naif. I don't know if that is true, though I hope so. Because if I've achieved what I set out to do, which was to share the euphoria (and the humiliations) of my learning curve and dejargonize the basic current thinking about microbes, then maybe the breakthroughs in microbiology that you read about in the papers will be a bit more accessible and make a bit more sense. We are just starting to understand microbial life. The term *microbiome* wasn't even coined until 2001, by the molecular biologist Joshua Lederberg. There are a lot of discoveries on the horizon, and it's going to help to have a little microbial literacy under our belts. "The planet is half-owned by microbes," sighed the microbiologist Moselio Schaechter, in his rich Italian accent. "Which means, if we don't know something about microbes, we don't know half of ourselves."

He's right. For all our technology, we still don't know the half of it. For me, studying invisible life expanded my perspective of the world, and it broadened my definition of life to include organisms that were beyond my ken. But once I understood that these creatures invented living and, as a result, are implicated in every aspect of every life, it raised my own soggy consciousness. I glimpsed another world and found out I share it with everything else on Earth.

By looking into the unseen, my sight cleared.

CHAPTER

I

Six Ingredients Connect Us All

H ave you ever felt a visceral connection with the environment around you? I have many times. I feel it when I hunt for mushrooms in the woods, when at first I don't see any, and then I see one, and then suddenly I see dozens. I feel it when I see the bloodred of a maple tree's fall foliage in the colorless night. I feel connected to a grove of trees depending on where I am standing in it. I feel something like love when I ride the subway train and imagine that everyone is running lists through their heads but we are all hopefully going to eat dinner tonight, and then sleep, and then start again in the morning. I feel overwhelmed when I walk into my father's ancient New England garden, with the partially rotted but still fruit-bearing pear trees covered in the traffic of ants, the seething piles of compost, and the feral vegetables bursting from soil marled and black as tar, the garden buzzing with flying insects and popping with jumping ones, all of it enclosed in vine-covered fences, the space damp with breath.

I started my study of microbiology because I had a hunch that those intuitive feelings might be explained by the presence of microbes. Maybe the connection I felt between living things could be found on the tiniest level, and the reason why I felt it was because I couldn't see it. Maybe microbes connected soil to plants and plants to us and us to each other. Maybe microbial life even constituted a physical connection between all living and nonliving things.

It turns out, they do.

But my journey learning how microbes connect our lives was a bumpy one. It wasn't easy going back to school at 55. I didn't head to college right away like a sensible person. I had to fail at self-teaching first.

I started out reading books and scholarly articles. Unfortunately, it didn't take long before references to precise sequences of amino acids and mitochondrial metabolism made me sleepy and vulnerable to YouTube cat videos. I was a faithful follower of the microbiology blog *Small Things Considered,* which usually had juicy opening sentences like "The quest to live longer is not only the stuff of myths and legends, it is also the goal of some serious scientific research." That would get me reading, but then within a paragraph or less (usually less), I'd get bogged down by enzymes that degrade reactive oxygen species and heat shock proteins that remove misfolded proteins, and within minutes I was utterly lost.

I spent at least a year trying to make sense of this stuff on my own, slowly becoming more despondent. I could read for 2 hours, but then when I closed the book, I wouldn't be able to repeat a single concept. I felt like I was floating further and further away from my goal, like a raft adrift on a current that looked like it was headed toward an island but actually pulls you past the shore and salvation, back out to sea.

Next I tried an online course, pretty much the first one that came up on an internet search for "Biology 101," thinking, well, if I knew what a misfolded protein was, maybe I'd have better luck with the reading. The $70 class took me through the scientific method (deductive thinking, as in all men are mortal, Socrates is a man, therefore Socrates is mortal; and the more sketchy inductive reasoning, as in Socrates is a man, Socrates is mortal, therefore men are mortal) to the characteristics of living things: sensitivity, reproduction, energy use, self-regulation, the role of chemistry in biology . . . and it was all going fine. I was getting 100 percent on the little quizzes at the end of each article (probably because I wasn't using my memory but my notes); nonetheless, I still felt terrific when I received a canned email complimenting me on my work. I carried on, usually for an hour or so before dinner: the atomic scale, the hydrogen bond, the parts of the cell.

But I couldn't retain the information from section to section. Over and over again, I had to visit Wikipedia for definitions of things I had supposedly learned—the cellular oxidation of glucose, ATP, the Krebs cycle—and I would *sort* of get it, but then when I'd try to reopen the bio class window, I'd be sent back to the home page and have to log in again. And of course, I never remembered my pin because I was born before 1990, and then when I finally did get back to the right page, I found, to my dismay, that my understanding of the Krebs cycle had degenerated into a blurry familiarity that was just too vague to move forward in the lesson in any profound way.

That's why I went back to a brick-and-mortar school. I needed a blood-and-guts teacher. I didn't have the background to take biology on a graduate level, and the idea of getting another bachelor's degree, and taking French and volleyball and all the other stuff I'd need, just seemed too preposterous at my age. So I applied to a continuing education course of study called Ecology, Evolution and Environmental Biology at Columbia University in the fall of 2015.

I had received my bachelor's degree in English from Barnard, Columbia College's sister institution across the street. Columbia didn't take girls when I was a student, and Barnard has never taken boys, although in 2016 the school started accepting applications from "individuals who consistently live and identify as women." I figured I'd be on familiar turf, and pathetic as it may sound, the idea of attending an Ivy League school gave my ego a boost, just as it had when I was young. Applying was simple, except that I couldn't remember what year I'd graduated and, pretty mortified, had to call and ask. The young lady in the registrar's office was relaxed. "Don't be embarrassed," she said. "People who graduated in the late seventies and early eighties forget all the time."

There was definitely a part of me that feared going back. I mean, I was pretty sure that studying biology would help me articulate my feeling that life was a huge collaboration of microbes or something like that, but those high-brow desires were marred by lowdown worries about how I was going to handle the time commitment, how I was going to manage homework and tests, how weird it would be to walk in a crowd of fellow students who were decades younger than me. I had read that there is an increasing number of "re-entry students" (a designation that made me feel like the space shuttle), and they're mostly females. I knew as soon as I hit campus, I would be scanning the crowd for them: the graying, the balding, and the thickening midsections. But most of all, I was worried that I just couldn't hack biology.

I checked out a forum online, The Well-Trained Mind Community, where a significantly younger person than me was thinking of going back to school for a nursing degree and was nervous about the workload. What a lovely thread. A host of cheerleaders encouraged her with countless "Go for its!" and "You're never too old to learn!" And I wanted to feel the love, too. I vowed to remember Henry Ford, who said, "Anyone who stops learning is old," and to take the wisdom of Helen Hayes to heart.

"Age," she said, "is not important unless you're a cheese."

I made an appointment with my Columbia student advisor in mid-May. I was pretty confident about returning to my old campus, though I hadn't visited since 1984, but it turned out I didn't remember any of the buildings. Graduation had just wrapped up, and I had to negotiate a maze of barriers and bleachers, which were continually changing as workmen hustled to disassemble the stage, not to mention the obstacle course of parent-graduate photo ops, selfie-stick groups, and other lost people wandering around like me.

My advisor, Matt, was tall, handsome, and exquisitely sober. Talking to him, I felt like I had a hangover even though I didn't. I told him I wanted to learn about microbial symbiosis and I wasn't interested in statistics or primate evolution, and he said, well, you really can't study biology without studying chemistry, and he recommended a modest load of two courses, Chemistry and Environmental Biology. I gagged. Chemistry: the culprit behind the abandonment of so many science majors. (As one friend of mine said, "Asked to perform a chemistry equation at the blackboard, I burst into tears and accused the poor, frightened TA of picking on me because he KNEW I couldn't do it. I fled the classroom and called my dad from a pay phone and told him I had to go home. That's how I went on to get a BFA in drama.") I only knew two types of chemists when I was in college—those that were so into the language of chemistry they could hardly communicate in English and as a result I didn't really know them at all, and those who made MDMA in the campus lab at night. I told Matt I'd rather not take chemistry.

And then he took out his big guns and suggested that if I felt that way, maybe I would be better off with an online course.

Of course, I signed up for chemistry.

Matt was unable to register me on his computer, so he sent me on to the registrar, where a young man patiently searched for my existence. He finally lifted his eyes from the screen and I think took his first good look at me.

"Are you an alumnus?"

"Yes," I said.

"What is your maiden name?" And then I popped up on his system, the name of the girl I was 35 years before.

"If you are okay with it, I suggest you just leave it. It's a hassle to change. But the registrar's office can do it if it matters to you."

It kind of did matter. I've been living under my married name for 25 years. The bulk of my professional life had been practiced under my married name. It was as if Columbia had just wiped away everything I had made of myself, everything, that is, except for the experience of having once been a student there. I didn't want to be know-nothing Eugenia Giobbi again.

"This is infantilizing," I told the registrar, and he laughed.

"Yeah," he said. "College is like that."

I was already tired of finding offices and waiting for someone to look up from their computer to answer my questions, and so I agreed to remain, in Columbia's eyes anyway, my father's child.

It felt like in one afternoon, I had rewound to age 20, and all of my experience and professionalism and skills never happened. But likewise, my prejudices and assumptions were irrelevant, too. I was a student again.

Matt had recommended I take a look at a biology textbook just to acquaint myself with some of the principles, which I did. Over the summer I worked assiduously, puzzling through the same stuff I'd puzzled through online. By the first day of class, I'd read 74 pages. Out of 1,200.

The moment I walked into Chemistry 101 and found my seat, I recognized this was not my tribe. My tribe is in search of the perfect neck pillow; my tribe is struggling to save at the same pace as the rise of college tuitions; my tribe takes 10 minutes to explain to a waiter how they want their three-ingredient martini made; my tribe is worrying over elderly parents and plans every wardrobe choice around the possibility of a hot flash. The undergrads around me were of a different ilk altogether, smelling of youth and blue jeans, their libidos as sharp as their minds.

Even the professor was younger than me. A 40-something fellow in a plaid work shirt and tie, our chemistry teacher waited, his legs apart in a stance so solid that if there had been an earthquake and the building shook, I think he'd still be standing, watching as we filled the seats in the wood-paneled lecture hall, his vibe practical, functional, rational. Two hundred of us settled down, all different colors and sizes and shapes, almost all in T-shirts, everyone looking at everyone else, though I imagined they avoided looking at me. I had felt for sure there would be someone my age in the class, but I was probably 25 years older than the next-oldest student. The gal that sat next to me had long blond hair and wore a teeny pair of shorts. She kept flipping her hair off her shoulder, and after it had splashed across my cheek a few times, I moved a seat over. She was very cute and looked to me like someone who had an active social life, and I thought, well, if she can manage chemistry and God knows what else she is taking, then maybe I can handle it, too.

The course was incredibly organized. There were no homework assignments and no mysteries as to when there would be a quiz, though we had to read from a textbook and answer questions on an online program called OWL, which sounded like homework to me. Additionally, we were expected to participate in an online chemistry forum called Piazza, where our interactions would be quantified in our final grade. And have we signed up for recitation yet? A few students raised their hands, mainly with questions about whether he would be grading on a curve and the percentage values of quizzes, but no one asked the question I was too embarrassed to ask: What is a recitation? Did everyone else in the class know what this meant? They must. I had a hot flash.

"Chemistry," said our teacher, "is based on measurements versus myth." He showed us some slides of alchemists in their workshops, puzzling over their flasks and fires, trying to convert base metal into gold, lobbing a few visual softballs before the real slides began—the periodic table, the weights of elements, the oxides and dioxides, the atom and its subatomic parts, the six elements of life: oxygen, carbon, hydrogen, nitrogen, phosphorus, sulfur. And then he whipped through the composition of the atom—its definition by number of protons, its isotopes based on number of neutrons, its allotropic phases: liquid, gas, solid . . . all of it just words I wrote down that represented concepts I was unfamiliar with. As I became increasingly confused, scribbling notes that didn't reflect any understanding, he pointed out humans are FULL of allotropes! As if bringing the subject round to one's self would somehow clarify everything. This was like a nightmare I still have, where I show up in a class and realize I haven't done any of the work. The only difference between chemistry class and my nightmare was the fact that I wasn't naked.

I looked over at the blond girl in the hot pants, half expecting her to be texting emojis to a friend. She had her laptop out with multiple windows open and was logging on to the online forum Piazza with her left hand while taking notes with her right hand, periodically flipping her shiny hair and casting eyes about the room. That's when I realized I was in trouble.

Education in the 21st century is very different from when I was a student. Indeed, the era of xeroxed handouts and blackboards and a student mailbox seems quaint, as if I came from the ranks of citizen botanists gathering seed collections in fancy dress. I worried that if I couldn't handle the basic technology of the class, what chance did I have with the homework? For me, multitasking was switching from my reading glasses to my distance glasses to see the PowerPoint screen. It wasn't that long ago when multitasking was thought to lead to poor performance. A friend of mine can't understand how her daughter, an excellent math student, does her homework more efficiently with *Game of Thrones* on her laptop. But a study presented at the American Academy of Pediatrics Conference & Exhibition in 2014 suggested that multitasking students had enhanced memory function and performance. Columbia's internet-friendly teaching model, in this class anyway, seemed tailored toward the innate abilities of digital natives. In other words, not me.

Life, our teacher announced, is a product of cosmic processes. The universe we observe today is thought to have started 13.8 billion years ago, when an inconceivably dense, hot, and unstable celestial, um, *state of things* detonated (Bang!) and expanded, and in the first second diffused and cooled into the tiniest particles, much smaller than an atom. Those tiny particles collided and merged into more substantial particles, which then collided and merged again to create the light-weight atoms of hydrogen and helium. Vast clouds of hydrogen and helium atoms collapsed or clumped into stars where heavier elements were formed. Two helium atoms make a beryllium atom. A beryllium atom and a helium atom make a carbon atom, the chemical basis of all life. A carbon atom and a helium atom make an oxygen atom, and so on. Ninety-four species of atoms are formed by these fusions, with iron being the heaviest element produced in the core of a living star.

There's a saying: Iron kills massive stars. It doesn't actually kill stars, but heavy elements do accumulate as stars age, and over time the core may become so massive it can no longer sustain itself and finally succumbs, producing a tremendous single explosion, a supernova, that blows debris at a speed up to 25,000 miles per second. That huge outlay of energy produces lots of elements, some even heavier than iron, like gold. "Think of all the gold in the world," explained the astronomer Deno Stelter in an email exchange. "Each gold atom was produced when a massive star ended its life in a supernova."

A supernova 4.6 billion years ago marks the beginning of our solar system. Its swirling dust and gases were pressed by gravity into a giant spinning disk that pulled gas and debris into its center, forming our sun. Dust that didn't get pulled into the sun swirled around, and as it did, the gases and material condensed into grains, and grains aggregated into planetesimals as small as a poppy seed or as large as Key West. Planetesimals smashed into each other to form protoplanets big enough to have gravitational fields, and they swept up the neighborhood, acquiring astro rubbish like tumbleweeds, growing into protoplanets that collided to become planets. Earth still orbits the sun in the same counterclockwise direction as the original disk of debris from which it was made.

A Mars-size body crashed into the proto-Earth, its molten core merged with ours, and the fragments flew back into space to eventually coalesce into our moon. Meteorites and asteroids composed of crystallized minerals battered Earth with the explosive power of nuclear bombs, which melted into the planet's expanding waistline, along with a shower of icy comets and their payload of water. As temperatures decreased, the planetary iron sunk to the core and the magma surface solidified. Light gases like helium floated away, but gravity held the heavier gases like carbon dioxide in place around the planet.

Our professor described Earth's position in the solar system as a very lucky break for humankind. Our planet is not too large and not too small, not too hot and not too cool, not too sunny and not too dark. Earth is just right for life. Biologists like to say life is a stochastic process, which means how it happens is totally random. But from what I can tell, so was the birth of our planet.

And then the class was over, and I filed out with 200 others, all of us in our own solar systems, some I imagined feeling like the sun in the center of theirs, but not me. I felt more like Pluto, which isn't even considered a planet anymore.

While I waited in line to pick up my student ID between my chemistry and biology classes, I looked up *recitation* on my phone, easy to do inconspicuously as everyone in line was looking at their phones, too. "A class period, especially in association with, and for review of, a lecture." Really? I had to take another class? And then my turn came up, and I gathered my ID: Eugenia Giobbi. The name shouldn't have surprised me, but it did.

The environmental biology course I'd signed up for was held on the fifth floor of a solid brick building with a worn marble lobby where countless sneakered feet had passed. I waited for a good 10 minutes with a handful of kids, mostly students on crutches, for a creaky elevator. The biology lecture hall was significantly smaller than chemistry's, with seating for maybe 50, and big windows facing west. The class filled up with students, again, all decades younger than I; beautiful girls, beautiful in their soft youth and spotless arms, and boys with long legs that they tried to fold under the small desk chairs like awkward origami. The room was cozier than the chemistry hall, and I couldn't help but feel the class was more genteel because of it. I sat in the third row.

Our professor, Shahid Naeem, was tall and gangly, with a dangly silver earring and a jaunty military cap; he walked back and forth in front of the class, his gait loose, a smile on the edge of every sentence. I'd looked him up. His lab, whose motto is Ecology with No Apology, introduced the team in a mock–*Law*

& Order video with a group photo at the end showing the scientists, who study birds and grasses and insects, as arm-crossing, ass-kicking bio-cops in black and gray suits and sensible heels.

The lab studies the ecological consequences of declining biodiversity: how the loss of a bird species might impact an ecosystem, how herbivores like worms respond to the loss of the birds, and how plants respond to changes in worm abundance. Other labs do this sort of work, too, but Professor Naeem's goes further. They look at the plant's inputs into soil, through rotting leaves and seeping roots that attract or repel microbes, and how they affect soil organisms like fungi, and further still, to how changes in microbial communities affect nutrient and energy cycling. In short, the lab looks at how the loss of something like chickadees might change everything: life, soil, air.

During the first class in environmental biology, we got the lowdown on the program, the expectations, and the players. We were shown a cover of *Biology* 10th edition, the textbook we were to buy, and Professor Naeem pointed out that we could get it from the Columbia bookstore. But that store, with its racks of notebooks and diploma frames, baby onesies, and hoodies emblazoned with the school football team logo ("Gear Up for Gameday!" said a bookstore email blast shortly before Columbia would lose its 24th football game in a row), was owned by Barnes & Noble. "To even the field," he recommended we buy it at the local indie, Book Culture.

Everything we needed to know was presented. A slide showed us the email addresses of the teacher's assistants, two graduate students who I noticed sat together in the front row. Even though they looked the same age as the rest of the class, smooth-cheeked and sneakered, they seemed more confident, like commuters who know exactly which end of the train platform to exit. We were informed of their recitation hours and shown graphs proving that in the past 3 years 27 to 37 percent of students got an A, 44 to 57 percent got a B, 13 to 24 percent got a C, and none got a D, though in 2012 3 percent failed. They were probably students who were unintentionally enrolled. We were to be graded on a sliding curve—"10 percent of all students will definitely get an A," said Professor Naeem—and I was relieved by this news, though I told myself grades didn't matter. It's not like I was planning on going to medical school. A friend and former professor at Columbia explained that no one at Columbia fails: "It's a brand thing," she said. I might have been a lot more judgmental about this were I not a student myself.

And finally, the syllabus. For today's lecture we were to have completed Chapters 1 through 3. This was news to me. The first day of class, and I was already behind. Professor Naeem whipped through similar origins-of-the-solar-system-type material as my first chemistry class had, and I wondered if every science class at Columbia started with the big bang. Plus he covered quantum mechanics, the nuclide chart, mature bulges (not what it sounds like), and the inside-out quenching phase that took place three billion years after the big bang.

"But I know you've had all this in high school," he said, "which is why I assigned the first three chapters for today."

Just for the record, if I had any of this in high school, I didn't remember a damn thing. Nor did we have to do the kind of study prep for tests like kids do today; in the late 1970s, we were discouraged from studying for the SATs at all, which were considered a sort of aptitude test back then, and anyway, my main concern at the time was not quarks and bosons, but whether my bosoms were going to get any bigger.

Then Professor Naeem asked one of those questions that seem simple but you know is not simple at all. What is life? I froze. I felt like I should give it a try, if only to encourage others. But I was afraid to bring attention to myself and be wrong, and so I looked down at my notepad.

The class struggled. One student said life is something that can reproduce itself. Another student said life is a cell. I racked my brain for a simple answer that described the vitality I felt all around me, not just in the churning campus, not just in the young Frenchman who was sending charged signals to a pretty just-showered athlete in class, but in the itchy spot between my toes, and on the surface of my desk, and in the dust motes that swirled in the light from the window. But I kept my head low.

"Life," Professor Naeem said, "is an adaptively evolving chemical epiphenomenon in which an organic matter system autocatalyzes the product of itself, and in so doing, consumes energy and increases entropy."

I thought he must be kidding. I'd come to school for clarity, not this tortured description that qualified every word and in the process utterly sucked out the possibility of any poetry. But he wasn't kidding.

When scientists look for life on other planets, they are looking for carbon-based life, because here on Earth, all life is based on carbon. It's an atom so versatile it can link in countless ways to other atoms. That day, Professor Naeem explained the foundation of an "organic matter system," like an armadillo or a willow tree. Carbon combines in different ways with oxygen, hydrogen, and nitrogen to make the complex molecules from which life is composed: proteins, lipids (fats), nucleic acids, and carbohydrates.

"So how did that happen?" asked Professor Naeem. How did carbon, hydrogen, and oxygen atoms end up as a carbohydrate, which ends up as your baked potato? Maybe the atmosphere cooked the atoms. In 1953 two American chemists, Stanley Miller and Harold Urey, demonstrated how geochemistry could give rise to biochemistry. "Harold Urey was a professor at Columbia," said Professor Naeem, pointing to a picture of a bespectacled fellow on the screen. "He petitioned for a higher salary, but when he was refused, he went to Chicago instead, where he did this study. Oops." Miller and Urey exposed an oxygen-free atmosphere composed of hydrogen, ammonia, and methane—similar to the atmospheres found on Jupiter, Neptune, and Uranus, the big gassy planets—to boiling water (to simulate hot ancient oceans) and to sparks of electricity to simulate frequent lightning. Within a week, they found that 15 percent of the carbon part of the methane gas converted into simple molecules that could go on to be more complex molecules, which are necessary for life. What's most important here is the carbon changed from an inorganic state as part of methane gas to an organic state as part of an amino acid, a critical ingredient that living things must have. Years later a box of data was found in Dr. Miller's lab that showed he'd conducted similar experiments with a variety of model primitive atmospheres. More than 30 different biologically important carbon compounds have been synthesized using the Miller-Urey technique. The recipe, it seems, is adaptable. Dr. Miller believed his experiment was definitive. "I happen to think prebiotic synthesis [and he means this coming together of atmospheric molecules] happened on the Earth," he said in an interview in 1996. "But I admit I could be wrong."

Indeed, some people—not my professor and not the authors of my biology book—think organic compounds got to Earth from outer space. (This is, by the way, a hypothesis for the origin of lots of things, like fungal spores; the tardigrade, a microscopic animal; and the pyramids of Giza—when in doubt . . . aliens did it.) That's not the same notion as Carl Sagan's, who said, "We are made of starstuff." Sagan was referring to the fact that life occurred from the elements formed by the big bang and subsequent smaller bangs like supernovas. No, I am talking about the idea that on another planet the Miller-Urey phenomenon or something else happened, and then the actual organic molecules that jump-started life traveled to Earth on meteorites like interstellar seeds. When the rocky, carbon-based Tagish Lake meteorite fell in a remote part of Canada in 2000, it contained carbon and an assortment of organic matter like amino acids, the building blocks of proteins. That got a lot of scientists very excited. (You can buy samples of the meteorite on eBay; $25 will get you a bit that looks like a crumb of burnt toast.) The 4-billion-year-old Allan Hills meteorite, which was found in Antarctica in 1984 and might have blown off the surface of Mars 17 million years ago, got a lot of publicity because biologists thought it contained some bacteria-like structures that look like very tiny, short strings of love beads. But as the paleontologist Paul Olsen pointed out in a lecture he gave at Columbia, "The human mind always wants to find meaningful patterns. It's like finding the face of Jesus in a pizza."*

But you can't pour the complex molecules of life into a cup and add water, like making Sanka. No scientist has made a cell from Miller-Urey's soup. But *something* compelling happened that led to the animation of these fats and carbs and proteins and nucleic acids. Life hasn't been reproduced in a lab because we aren't seeing something; there's some missing circumstance that we haven't identified yet, that needs to be added (or taken away). Was the animation of life a one-time, fortuitous thing? It's hard to know. Maybe every second molecules are coming together in the wild to make proteins and fats, but there's no time to witness what it is that makes them transition into life because as soon as these macromolecules appear, they're degraded or eaten. The origin of life, like the big bang, is still a secret. Professor Naeem put a Gary "the patron saint of scientists" Larson cartoon on the screen, where two PhDs are standing in front of a blackboard and on the left are all these calculations and on the right are even more calculations, but between them it says, "Then a miracle

---

*Some recent international entries in Jesus's face foods: Jesus's face on a burnt fish stick, a breakfast taco, the spots of a banana peel, naan, a grilled cheese sandwich, and a fried pierogi.

occurs . . . " It was his way of telling us that however the molecules of life formed, they are not in themselves life.

So while we know how chemicals synthesize into the stuff of life, at the moment, there is no consensus as to what made those macromolecules collaborate to make little organs with special functions, what made them, in short, *live*. The central dogma of biology says that DNA (deoxyribonucleic acid)—coded genetic information—is transcribed into little cassettes called RNA (ribonucleic acid) that each program for the synthesis of a protein. Imagine DNA as an encyclopedia. RNA would be the entry on how to build a bicycle, and the protein would be the bike. It turns out RNA, the "how to build a bicycle" entry, can make copies of itself, and it can build the bike, too. Because RNA can self-assemble and reproduce, the most plausible hypothesis for the origin of life is it all started with nucleic acid.

This was annoying. Here I was, starting a biology class, and the experts were already admitting that the most fundamental question remained unanswered. I'd signed up expecting all the answers I wanted to be available, like room service in a really good hotel. But knowledge, like organisms, evolves. The strongest ideas survive and the weakest ones go the way of the dodo; the whole picture of life isn't complete. First we learned what existed before the big bang is unknown, and then we learned how complex molecules formed life is unknown. But we did learn that for whatever reason and by whatever means, those complex molecules—proteins, nucleic acids, carbohydrates, and lipids—began making things. And one of those things was a membrane.

"Membranes," said Professor Naeem, "are a very big deal." The first cell was encased in a lipid, something like a bubble of oil that did not dissolve in water. In an ocean full of floating macromolecules, this little bubble of fat created a profound reality: an inside versus an outside, the ability to be separate from the environment, the potential to enclose stuff. Life as we define it started when the working parts for self-sufficiency and reproduction became contained, and there was self and nonself.

Membranes became more complicated over time, adding layers and properties, and so did the stuff they contained, but the basic system was an innovation that lasted. To this day, all cells, including those that make my body, are protected by a layer of water-repelling fat. I glanced around the classroom

chuckling, looking for someone else who was struck by the irony that the border between self and nonself was fat, but I had no takers, maybe because we were at the end of class and my fellow students already had their cell phones out.

I had many revelations during the process of studying biology. While cooking dinner that night, I had my first. All living things, from a bacterium squirming in a raindrop to the cactus in my dentist's waiting room to the president of the United States, are just different recipes cooked from the same limited number of ingredients, and when we die, we break back down into those ingredients. Then those ingredients are shuffled around and cooked into something else, or released into the atmosphere to be utilized later. This was the notion of the Greek philosopher Epicurus: There is a limited amount of matter on Earth, and it keeps recycling over and over in different living and nonliving forms. His fellow Greek, Heraclitus, put a slightly different spin on it, which also summed up my first day back in college: Everything changes and nothing remains still . . . you can't step into the same stream twice.

So on the most elemental level, my body is a renter of matter that has been elsewhere before. I am connected to the past and the future by six primary ingredients. I think we all want to feel a part of something bigger than ourselves. I just didn't realize we already are.

How Microbes Created
the Air We Breathe

M y chemistry and biology textbooks together cost $240 and weighed 14 pounds. That may be within the American Occupational Therapy Association's guideline of no more than 10 percent of one's body weight for a safe book bag load, but they seemed really heavy to me. My brother Cham had to take his 14-year-old son to the pediatrician for a suspected case of book bag disc crush. "I told him, just take home the books you need," said Cham. "But boys aren't like girls who think, what do I need today? Boys think, if I've got everything, how can I go wrong?" My nephew turned out to be fine, but since I had no intention of joining the legions of people my age who can rattle off the region and number of their injured vertebrae with the expertise of a chiropractor, I decided to do all my reading at home.

The chemistry textbook was packaged with a *Survival Guide for General Chemistry* that featured daunting chapter heads like Chemical Reaction Stoichiometry and Gas Phase Equilibria. Also included were *Essential Algebra for Chemistry Students*, which I agreed would be essential for me, and instructions for logging onto OWL, the online learning system where we were supposed to complete practice assignments.

I spent a frustrating hour figuring out how to get onto the OWL site as I encountered the usual array of missteps: registration, password, the scribbling of my password on the back of an envelope, the immediate forgetting of my password, the scramble to find the scrap of paper I wrote the password on, then saying screw it and trying my standard passwords and after a few rejections the system threatening to block my access altogether. By the time I'd logged onto the chat group, there were already conversations going on among students who had completed the first chapter of the book.

"In one of the questions on OWL, the problem said that the limiting reactant has the lowest mole available/coefficient in the equation. I understand that the limiting reactant has the lowest mole available, but does that also mean its coefficient is always the smallest in a balanced equation?"

I felt a spark of panic. I looked up *limiting reactant*. I looked up *coefficient*. I looked up *mole*, too, but the definition just had me more confused: The amount of pure substance containing the same number of chemical units as there are atoms in exactly 12 grams of carbon 12? Is there anything more random sounding than that? I was beginning to get a queasy feeling that my advisor hadn't understood how steep my learning curve really was.

It didn't take 24 hours for me to decide to drop chemistry. When I next had class, I went straight to the registrar with my withdrawal form. I was embarrassed and felt like I had to explain myself, but the registrar didn't care. She never even made eye contact with me. Neither did the bookstore, where I went to return my 1,075-page textbook. I purchased it for $158 on a Tuesday, and I sold it back to Columbia that Thursday—for $24.

It was humiliating to concede defeat before I'd read one chapter. But the truth is I was ready to bail when I realized I had no idea how to operate the scientific calculator required for the course. I mean, what does Statvar and LN, COS and DRG mean? I couldn't even figure out which button to press to make it divide.

But once I decided to drop chemistry, I attacked the biology reading with compensatory vigor. This I was going to get. I may not have what it takes to comprehend dumbbell-shaped electron orbitals, but damn it, I was going to master stuff like seeds. And since I was most certainly not fresh out of high school biology or chemistry as Professor Naeem had assumed, I had to read the first three chapters of my biology textbook, which, to my dismay, contained more material than I had reviewed over the entire summer. This high school–level stuff included the molecular basis of life, the properties of water, the chemical building blocks of life, cell structure, and cell membranes, most of which, ironically, concerned chemistry.

I waded through pages and pages of diagrams depicting the nature of atoms and their corresponding electron orbits, the construction of chemical bonds, acids and bases and logs. I studied endless diagrams of the functional chemical groups, all ending in *xyl* or *ntyl*, as if the comfort of vowels was not allowed. The reading was more like deciphering. I had to constantly go back and reread the definition of an isomer, a dalton—and at one point, I realized that the diagrams of the molecules I was looking at weren't, in fact, flat, but three-dimensional or twisted like a licorice stick. Why couldn't the authors have mentioned that up front? Is it some kind of scientific *sin* to use an analogy every once in a while?

I took copious notes from the textbook, writing whole sentences directly from the text, and then spent hours translating them: "What occurs within the cell on receipt of a signal is signal transduction" eventually became "signal transduction is when you are frightened and the presence of the 'being frightened molecule' tells your cells to amp up the adrenaline so your heart can beat faster and you can run away."

On the other hand, there were many facts in my textbook that I could not translate in any meaningful way at all because I was outgunned by the jargon. "The effect of phosphorylation on cell function depends on the identity of the cell and the proteins that are phosphorylated." Unfortunately, with only a watery understanding of what phosphorylation is, I translated this as "the effect of something depends on the nature of the something and the thing that causes the effect." I began hoping a miracle would happen and I'd have some kind of aha moment in the lecture. It felt like when I used to pray for a snow day.

Nonetheless, between the first and second biology lectures, I reviewed 100 utterly dense pages, and each chapter was packed with questions that I felt compelled to try, like "Why do phospholipids form membranes while triglycerides form insoluble droplets?" I hacked my way through these thickets of material, choked with obstacles like a jungle, until finally I reached a kind of clearing where the sun was shining and the air was fresh, and I started to read about living things: the first organisms on Earth, how they managed to live, and what they did that changed Earth forever.

Earth's first organisms emerged about 3.8 billion years ago in a watery world. Ocean temperatures may have dropped to below boiling, but they were still hot. If a human took a swim from that ancient beach, he would have come out with third-degree burns. Artists' illustrations of this eon, the Archean, depict trippy red-and-purple methane skies over volcanic eruptions, 1,000-foot tides, a larger, closer moon, and a dimmer sun. The Earth's crust was thin, with molten rock near the surface. It seems like a pretty inhospitable setting for the emergence of life, which when I was young was portrayed as a fish with flippery feet crawling out of a mucky pool in the Everglades.

Earth is around 4.6 billion years old, but because the surface of the planet was molten for a billion years, the oldest rocks are about 3.9 billion years old. And those rocks, traveling on tectonic plates, were crushed and compressed and cooled and reheated over and over again. They endured too much transformation to retain any fossilized proof of what might have lived before. As a result, the earliest fossils we have, of single-celled organisms lacking a nucleus, come from the earliest rock from which fossils could be saved. That means when scientists date the appearance of life on Earth to 3.8 billion years ago, it's not because that is necessarily when life appeared. It's when we can prove the existence of life based on what we can find in the oldest rocks.

Back then, Earth's atmosphere would have been pretty nasty to us, but not to the organisms that evolved to exploit those conditions. All life-forms need water, food, and a way of extracting energy from that food. That's the system in place for all animals and every other creature, as well. The first life-forms made their own food by using an energy source like volcanic heat to cook sugar from otherwise inedible raw ingredients like CO_2 or hydrogen sulfide. It may seem like a pretty meager way to eke out a living, but that's what was available on ancient Earth.

All the diversity of life sprang from these first organisms, what my biology book called, rather disrespectfully, "the bubbles" that became living cells. These were living cells without a nucleus, called prokaryotes, from the Greek for "before" and "nut" or "kernel," because they are pre-animal, pre-fungal, pre-plant. But that's kind of a misnomer because prokaryotes never ceased traveling their own evolutionary track. Okay, some went on to become complex organisms like flamingos and supreme court judges, but the majority continues to exist in vast numbers. That's what bacteria are.

We had one lecture on prokaryotes. Considering that they spent a couple billion years perfecting how to live off abiotic (nonliving) resources, that all subsequent life evolved from these critters, and that they are still the most abundant life-forms on Earth, that seemed pretty skimpy to me. But this was an intro to biology class, which meant that a lot had to be covered in two semesters. In order to do that, Professor Naeem had to knock out a huge amount of material in 55 minutes. There were well over 60 slides, and he swept through them in one continuous monologue.

Modern prokaryotes are probably much the same as their forefathers: simple capsules containing a primary mass of DNA and one or more secondary DNA threads or rings called plasmids. The cell also contains a few more parts, like ribosomes which spit out proteins, and locomotion aids. Some prokaryotes drag themselves along using a little grappling hook. Others have a spinning tail that motors them around, relatively speaking; a bacterium swimming in a drop of water is relative to you swimming through a sea of molasses. They sample the chemicals in their surroundings, eventually testing enough of their immediate area to determine in which direction they might find food.* Prokaryotes can be as translucent as a contact lens, or pigmented, or glow in the dark, and they are shaped in countless variations of spheres, rods, and spirals. You've seen the quintessential bacteria in articles like "Your Keyboard: Dirtier Than a Toilet." Often gaudily colored to make them easier to distinguish, they look like furry DayGlo vitamins.

The vast majority of organisms on Earth are prokaryotes, and there are

---

*A 2010 study of indigenous Mexican mushroom foragers comparing male and female hunting styles found the men tended to cover great distances and expend a lot of energy (suggesting they are really programmed to follow game) while the women covered less distance but tended to methodically scope out the food possibilities. Bacteria are more like women foragers.

two types: bacteria and archaea. They look similar, so similar scientists lumped them in the same group for a long time. Both are found everywhere, from soil to oceans, mouths to guts, but they are as genetically different from each other as I am from a cactus; they differ in the composition of their cell walls and in their evolutionary relationships to other life-forms. Scientists have studied bacteria more than archaea, and some of what I learned about bacteria may eventually be found true of archaea. But, of course, maybe it won't.

Prokaryotes (bacteria and archaea) are miniscule. Hundreds of thousands of them can fit in the period at the end of this sentence. The geologist David R. Montgomery made this analogy: If a bacterium was the size of a baseball pitcher's mound, you would be the size of California. And the discovery of nanobacteria is revealing organisms that are way smaller than that. While some prokaryotes are big enough to be visible, just barely (for example, *Thiomargarita namibiensis*, a bacterium that lives in ocean sediments on the continental shelf of Namibia, is about the size of a bumblebee's eye), there is good reason for most bacteria and archaea staying small over the eons. They don't have mouths or anuses or stomachs. Nutrients, waste, body heat all diffuse in and out of the cell, and the smaller the surface area of the cell, the better diffusion works.

Prokaryotes don't have sex to share genes, either, not in the way you and I do. They can make a pilus, which, as a PhD student explained to me, is as close to a penis as a prokaryote will ever get; it's a kind of probing hollow hair that bridges two prokaryotes and allows genetic material to flow through. What they do is not even called sex. It's called conjugation, as in a link or connection, and it allows bacteria to acquire genes that help them do new things like resist a particular antibiotic. Here's an example of how effective this form of genetic transference is: Imagine you are lactose intolerant—you lack the gene necessary to make the enzyme that degrades lactose—but you get a job in Switzerland and every meal includes cheese. You have a problem. You really need to eat cheese if you are not going to starve. If you transferred genes the way prokaryotes did, the problem is solved when you encounter a Swiss cheese-eater and shake hands and the gene that allows him to digest lactose seeps via the handshake into you, and now you have the gene that codes for lactose tolerance. And you can eat cheese right away. Now let's add a weird twist. Different species of bacteria or archaea can share genes. So, imagine the Swiss cheese-eater is a dog whose paw you shook. You could still get, and be able to use, the dog's lactose-tolerant gene.

Prokaryotes can even uptake DNA directly from their environment. When a bacterium dies, its cell ruptures and fragments of its DNA spill into the immediate environment. Those fragments of DNA can be taken up by another bacterium in the vicinity. It's as if the Swiss cheese-eater sat on a chair and spontaneously combusted and left behind some genes and then you sat on the same chair and picked those genes up. But let's be clear. This is the random acquisition of potentially useful genes, so to continue with the analogy, if you shook hands with a Swiss cheese-eater, you might get his lactose-tolerant gene, but you could get a different gene altogether. You might become a gifted yodeler. Natural selection occurs when the genes allow the bacterium to survive a little longer, and so divide and multiply, and pass those genes along. (Natural selection suffers from the word *selection*. It sounds like there is intent at work. There isn't. It's more like natural acceptance. If you've got the genes to survive whatever environmental changes are being thrown at you, nature accepts your bid for another day on Earth.)

Prokaryotes also share genes by means of their primary predator, viruses. Viruses that infect bacteria are called bacteriophages—phages for short—and there are a lot of them. "If phages were the size of a beetle," the biologist Marisa Pedulla told the magazine *Science News*, "they would cover the Earth and be many miles deep." But not all phages kill, as bacteria have developed mechanisms to resist their effects. As a result, some phages pick up genes from one bacterium and carry it to the next. Again, this process is totally random, but if the gene improves the bacterium's survival, it gets passed along to the next generation.*

Prokaryotes increase by division, and they increase exponentially. *Escherichia*

---

*Some bacteria cope with viruses via a type of protein that identifies and destroys viral DNA. The first to be discovered is CRISPR/Cas9 (one part of the protein molecule identifies the viral DNA, and the other part takes it out); it is now being studied for its ability to remove disease-coding genes from humans. If a bacterium were a word processing program, CRISPR would be the cut-and-paste function. We have co-opted it from bacteria and are seeing how we can apply the function to humans or plants or fungi. There's already a CRISPR-edited mushroom. A plant pathologist at Pennsylvania State University has removed the gene that codes for the enzyme that causes browning. She made a white button mushroom that stays white despite its age. (The USDA passed on regulating it.) One of the teachers in my department said the discovery of CRISPR was the most exciting thing to happen in molecular biology in 50 years and someone was going to get the Nobel for it. It's also one of the most controversial because CRISPR can potentially remove any gene we want, and the removal of that gene is permanent. Your children can't inherit a gene that isn't there anymore. It's important to be sure that a gene we remove for one disease doesn't also code for resistance to something worse.

coli, which lives in your guts, divides about every 20 minutes. Given idealized conditions (which are implausible), after 36 hours one bacterium can produce enough offspring to cover the entire surface of Earth with a 1-foot-deep layer of *E. coli*. In 39 hours, the Earth would be 8 feet deep in bacteria. In less than 2 days, their mass would equal the planet's.

Scientists used to think bacteria were isolated; that when a bacterium divided, the two sisters didn't know about each other. But if bacteria or archaea find a nice food source like a bit of raisin stuck between your teeth, they will settle in and begin dividing, and the cells will stick together into a durable colony called a biofilm. A biofilm can be composed of billions and billions of cells. They can even grow into complex communities with hundreds of species present, adhering to one another and to every surface that has moisture and nutrients by means of a sugar-based goo, creating a matrix that can be as tough as shellac. But this isn't a simple conglomeration of cells. They aren't just stuck together. They work in unison to regulate the functioning of the whole colony. "Among the big-time adjustments in our views of the microbial world," said microbiologist Moselio Schaechter, "has been the recognition that bacteria do not particularly behave as individuals."

Biofilms are everywhere, or at least, everywhere with even the remotest possibility of life, but mostly where it's predictably wet. The slick stuff coating river rocks? Biofilm. Desert varnish, the reddish black veneer on the rocks in Utah's Canyonlands? Biofilm. The slimy gunk inside the sink drain? Biofilm. Your daughter's recurrent ear infection? Biofilm. It's likely your teeth are covered in biofilm right now (that's what dental plaque is), and they are probably all over your toothbrush bristles, too. Toilet rust? Biofilm. And if you look at your baby's pacifier under a microscope . . . well, don't.

If there was ever an example of strength in numbers, biofilm is it. One aimless *Salmonella paratyphi* bacterium on a subway seat isn't going to give you typhoid fever. But a lot of them in a biofilm can present virulence. That's because bacteria in biofilms can coordinate their behavior in more ways and for more purposes than an individual bacterium can. When there is a change in the immediate environment—like more or less food, or the presence of a competitor for that food, or a spritz of Lysol, or even an immune response if the bacteria are causing an infection in a person—then the microbes need to share what's going on in order to mount a response quickly enough to survive. To do this, bacteria use a kind of groupthink, a crowdsourcing mechanism invented many millennia before ours, called quorum sensing. It's how they talk to each other.

Bacteria in the same neighborhood need to reach a certain number—a quorum—in order to turn on the genes that help them adhere. Once in a biofilm, bacteria seem to exhibit behaviors that are analogous to animals, including humans. In fact, they may be the most cooperative domain of life on the planet. To illustrate this phenomenon, Professor Naeem showed us a slide of a very cute squid, the bobtail, also known as the dumpling squid; they are about as big as your thumb. (Throughout his lectures, Professor Naeem inserted pictures of animals like funny frogs and adorable lion cubs between the incomprehensible illustrations of things like chemical pathways. I recognized this as the subterfuge it was—a little breather to recapture our flagging interest. While these relief slides were grievously few and far between, I was, nonetheless, grateful.) Bobtail squid hide in the sand during the day and come out to hunt in shallow coastal waters at night. But when the moon is up, the squid casts a shadow on the ocean floor, which alerts predators. Every night, the squid attracts a bioluminescent species of bacteria, *Vibrio fischeri*, by luring them with yummy chemicals into a compartment near its belly. Once inside, the *V. fischeri*, which aren't bioluminescent as individual cells, find each other using chemical signals, and when they reach a quorum, they turn on the light and the squid's stomach glows like airplane landing lights, obscuring its shadow on the sand. Now the squid can carry on hunting with impunity. The takeaway is quorum sensing (finding each other) controls biofilm formation (enough numbers to make something happen).

Bacteria are pretty much all about looking for food. Once they find it, they can settle into a community, and as a community, they can exhibit really complex behaviors that are more like social animals. Most biofilms are composed of multiple colonies with one or hundreds of species (as well as all kinds of microscopic gunk, like particles of clay, silt, even blood, depending on where the biofilm is located) that compete for resources just like everything else in life. They use chemical weapons to ward off other microbes, and we've harvested some of those chemicals, like antibiotics, for our own use. They coordinate their behavior to dominate a threat or use population numbers to crowd out other species of bacteria. Under some conditions, a bacterium that is part of a clonal colony—meaning all the bacteria have the same genes—can become a suicide bomber, blowing itself up and releasing toxins that kill nearby bacteria from a different genetic group. But it's not all war. Biofilms pool nutrients, they take turns feeding when food is limited, and some even heal wounded compatriots by sharing healthy cell walls. A single bacterium, even though it might be able to sense trouble when the Lysol spray comes out, can't process its own information. The ability to process information and react is a group phenomenon.

Prokaryotes even die differently from the rest of us. A biofilm doesn't die instantly when exposed to Lysol. The time it takes to kill a biofilm is proportional to the size of the community, because while some are dying, others are multiplying by cell division. And rather than die, some species in the Firmicute phylum (of which *Clostridium botulinum,* the bacteria that causes botulism, is a member) can go into a kind of hibernation, forming a tough cell wall-cladding called an endospore that helps the cell resist environmental stresses. It's the hardiest known form of life on Earth, and bacteria can build this shell in a matter of hours. Then they wait until conditions improve before returning to their normal state, even if it takes 100,000 years. Or who knows? A million years.

We did read a chapter on viruses, and we learned some very basic stuff, but the lecture mainly focused on how viruses affect us: how HIV infection works, the horrors of bloody Ebola, how viruses can cause cancer. Viruses are strands of nucleic acids, the stuff of DNA or RNA, in a protein bodysuit. They look like drums, crystals, bombs, and mechanical spiders and are usually eight times smaller than a bacterium and way more numerous, like 10 times more than all of the other cells on Earth combined. They are really just a set of instructions— rogue DNA. Once introduced in a host, those instructions take over the machinery of the host cell to build copies of the virus and commence with whatever mischief or aid they are programmed to do. But how they evolved, how they occupy all possible environmental niches, whether they were present at the origins of life, and what roles they play are all questions that are unanswered but someday may very well change our understanding of life.

Whether or not viruses are alive depends on what you consider alive, but on a scale of aliveness, they are pretty low down. I don't think viruses measure up to Professor Naeem's definition of life from the first class. But interestingly, viruses preceded all other life-forms in the Diversity of Life on Earth section of my biology book. It's almost a passive way of communicating that, yes, viruses are alive, and they are key to diversity on Earth, but we haven't figured out how. When it comes to viruses, we're mostly clueless. To that end, I did an informal survey of dozens of scientists at an American Society for Microbiology conference in Boston, asking if viruses should be considered alive. Across the board, they told me the problem with determining whether viruses are alive or not lies in the limitations of our definition of life.

Bacteria and archaea are the oldest, structurally simplest, and most abundant forms of life as life is currently described. The total amount of carbon contained in all prokaryotes is about equal to all plants, including the forests in the Amazon, the taiga, and the Congo, give or take. The number of individuals is as difficult to comprehend as their minute size, somewhere around five followed by 30 zeros (here goes: 5,000,000,000,000,000,000,000,000,000,000), more than there are stars in the universe. Astronomers don't know what constitutes most of the universe. They label this enigma dark energy and dark matter. Bacteria and archaea are the biological equivalent. The vast majority have not been characterized. They are so intangible that biologists call their collective mystery microbial dark matter. They are the living analogy of the universe.

Professor Naeem explained to us that the first life-forms survived on ingredients like atmospheric gases and minerals. When I realized that means the elements crucial to life are cycled into the food chain by the intervention of microbes, I thought I was going to fall out of my chair. For me, it was a mental breakthrough. Prokaryotes are the link between the living and nonliving worlds. The food chain originates when elements move from inorganic to organic states, from a gas in the atmosphere to carbon in a cell, and prokaryotes do that job. They are the foundation of the food pyramid that the rest of us depend on.

They do that job because that's how many prokaryotes feed, and how they always have fed themselves. Indeed, there are single-celled microbes out there making a living in environments today that are similar to what scientists think primitive Earth, or parts of it anyway, looked like: toxic gases, intense saltiness, vicious heat. When you look at those microbes, you may very well be glimpsing what the earliest forms of life looked like. In fact, the chemistry of the water around deep ocean hydrothermal vents is much like the chemistry of early Earth, and there is evidence that we may descend from the microbes still found living in these ferocious environments.

Relics of the ancient world, these microbes are called extremophiles, and they live in the most forbidding nooks and crannies of the planet. Of course, these creatures only seem extreme because the conditions they've evolved to live in would kill us. It's like the mycologist Nicholas Money wrote in *The Amoeba in the Room*: "Everything in biology is somebody's extremophile." I read about acidophiles, acid-loving microbes that are happiest in environments 1,000 times more acidic than battery acid, and halophilic, or salt-loving, microbes that thrive in places like salt crystallization pools, which are 10 times saltier than seawater. Barophilic bacteria, the pressure-lovers, live in sea geysers that spew a cruel mix of 572°F seawater, hydrogen sulfide, and magma under a crushing pressure of 8 tons per square inch. And hyperthermophiles, the heat-loving microbes, prefer temperatures hotter than boiling water. Life, noted Thomas Brock, the scientist who carried out key research on Yellowstone National Park's hyperthermophilic microbes, is possible at *any* temperature as long as there is liquid water.

Even below the seabed, microbes thrive. Researchers found cryptoendoliths, microbes that colonize little cavities in porous rock, living off methane and sulfur almost a mile *under* the seabed. Scientists drilling below the Juan de Fuca Ridge off the west coast of the United States said they found them no matter how far down they drilled. Since most of Earth's crust lies beneath the oceans, the discovery of these rock-abiding microbes suggests Earth's crust might be one huge habitat, and should similar microbes be found throughout it, the crust "would be the first major ecosystem on Earth to run on chemical energy rather than sunlight," said the ecologist Mark Lever, who led the study, in an interview with *Nature* magazine.

There are other extremophiles, too—subsets of the endolithic microbes, those that live in coral, animal shells, or the pores between mineral grains of rock, and microorganisms that live on permanent snow, called psychrophiles (*psychro* means "cold"). There are microbes—both bacteria and fungi—thriving in the radioactive environment of the damaged nuclear reactors in Chernobyl and microbes living in the liquid asphalt of the La Brea Tar Pits in Los Angeles, burping methane. There is a microbe adapted to survive in every condition on Earth and even off it. According to studies published in *Astrobiology*, bacteria in the genus *Gloeocapsa* survived for 18 months in space before they were brought home.

These are hardy critters, "the world's most resistant and persistent biological entities," wrote John Ingraham in *March of the Microbes*. Indeed, a colony of microbes found living in a deep brine lake under the dour McMurdo Dry Valleys in Antarctica turned out to be well over a million years old. They break down iron in the bedrock for their energy needs really, really slowly. It turns out they can survive eight million years in a frozen state, though once resuscitated they are pretty beat, like an ice pop that's been hiding in the freezer too long.

For a billion years prokaryotes lived by means of inventive survival strategies, like making meals from carbon dioxide, water, and dissolved iron. But all that changed, everything changed, when a type of ocean-living bacteria started making its food by using the energy of the sun to grab carbon from CO_2 and to break water molecules and use the energy binding them to turn that carbon into sugar, which freed oxygen in the process.

Those bacteria were and still are cyanobacteria. Cyanobacteria are a large group of different kinds of bacteria that get their energy from photosynthesis, like plants, but they appeared on Earth way before plants did. Today they live in every body of water on the planet clear enough to let in light. Through a microscope they look like green lozenges, and their colonies look like necklaces of jade beads or strips of impossibly small green movie film. With the naked eye, it's possible to see these beads and films linked in groups that look like pea soup floating on top of the water. You can even see them blooming from space. They look like eddies of pale green that resemble a swirling tropical storm. There are more than a billion metric tons of these tiny creatures on Earth, an estimated one thousand trillion trillion or more cells floating in the seas.

I hit a few bumps in class at this point. I always thought respiration was breathing. It's not. Breathing is how we get gases in and out of us. Respiration is when a cell combines oxygen and glucose (aka food) to produce usable energy. That was a manageable mental realignment. But then we were taught that, no, photosynthesis wasn't just when the energy of the sun was used to remove the carbon from CO_2 and an oxygen molecule escaped in the process. Nope. I had to unlearn this and relearn that photosynthesis was more complicated, with multiple parts. There is a stage where carbon is turned into glucose, the sugar plants use for energy, and a stage where the plant converts solar energy into chemical energy. Fine. But the energy stage consists of two parts, photosystem I and photosystem II, and photosystem II actually comes first. It's when a cyanobacterium uses solar energy to break a water molecule, H_2O, into hydrogen and oxygen, and *that* oxygen atom is the one released into the atmosphere, not the oxygen in CO_2, which is used for something else. Despite the fact that photosystem II happens before photosystem I, it is numbered two because it was discovered later, and the scientific community made this terminology decision in order to, yep, avoid confusion.

For every water molecule split by an ancient cyanobacterium two billion years ago, an oxygen molecule was released into the atmosphere. And there were plenty of water molecules to go around. Every 10 drops of water has about the same number of molecules as there are stars in the universe. This system produced so much food that the cyanobacteria multiplied like crazy, which led to more splitting of water molecules, which led to more oxygen being released into the atmosphere. These bacteria freed oxygen that was shackled to carbon and set it loose on the world.

The results were devastating, but they led to us.

At first, the massive amounts of oxygen being pumped into the atmosphere by cyanobacteria were captured by dissolved iron in the oceans. And there was lots of iron around to be oxidized. By mass, iron is the most common element on Earth. The core of the planet, both the solid inner core and the molten outer core, are mainly composed of iron. Iron exists in a range of states, but the most common are ferrous and ferric. Ferrous iron, sometimes called fresh iron, is the more bioaccessible form. Because ferrous iron is reactive to oxygen, when cyanobacteria oxygenated the earth, the oxygen converted the ferrous iron into the biologically less available form of ferric iron, or rust.*

Iron is vital for a wide variety of processes in almost all organisms, from DNA synthesis to the transport of oxygen via red blood cells. (Iron is an important component of hemoglobin, the protein in red blood cells that binds to oxygen. It's the iron interacting with oxygen that makes blood red.) Ever since the atmosphere became oxygenated, ferrous iron has become scarce. But since we really need it, we hoard 2 to 4 grams in our bodies; we don't even have a method for excreting it. There is a disorder called pica that is characterized by an appetite for nonnutritive substances like metal and chalk and is commonly seen in small children and pregnant women—it is often associated with some kind of vitamin or mineral deficiency. My brother-in-law Paul was a rust-eater as a child. "It tasted like licorice and dust and cobwebs," he said. "I chewed it like tobacco." We all have an ingrained appetite for ferrous iron, though Paul's passion for rust was a misplaced desire, and one, I am happy to report, long since quenched.

Bacteria need ferrous iron, too, and they get it any way they can, even if that means stealing yours. Indeed, if someone injected you with ferrous iron, you would experience more infections because you'd become more attractive to bacteria. Ferrous iron is like a bacteria magnet. And the more iron-deficient you are, the less likely you are to attract the pathogenic bacteria that are after it. To this day, most of the oxygen on Earth is bound up in rusted iron in the planet's mantel.

But a point came where there was simply more oxygen than could be sequestered in iron or elsewhere, and the excess free oxygen started to accumulate in the atmosphere. The ancient cyanobacteria assembled into biofilms, which layered into microbial mats. The mats captured silt and sediments and they eventually built firm, hassocklike structures called stromatolites (or layered rock), kind of like microbial Napoleon pastries that settled on the floor of shallow, sunlight-drenched seas. They are rare but still exist in the chemical-rich waters of Shark Bay, Australia—gray rocklike loaves that grow by adding a microbial mat every year like a fresh layer of paint. Stromatolites and their sister microbial structures, thrombolites (which grow in clots rather than layers), are atmosphere machines, and over "a mere 300 million years," said

---

*It should be said that types of bacteria converted ferrous into ferric iron before the presence of oxygen, but their impact on the overall ferrous-to-ferric-iron ratio wasn't nearly as great as when the atmosphere filled with oxygen and rusted everything.

Professor Naeem, they changed the planet, and subsequently the course of evolution. Even today, though Earth is covered in photosynthesizing land plants and the seas are full of photosynthesizing phytoplankton, bacteria floating in the oceans produce a fifth of all the oxygen in our atmosphere.

Some of the oxygen interacted with ultraviolet radiation (UV) from the sun to form ozone. You've smelled it before the rain. When lightning strikes, it starts a chain reaction that can produce ozone, which has a scent that is sweet and sharp and clean, like bleach. Ozone is a protective layer of gas that decreases UV radiation on land, and consequently reduces the chance of potentially fatal UV-related DNA mutation. In fact, every cell that lives today—and that includes the cells that make us—contains an ancient gene inherited from bacteria that codes for UV repair, a gene that predates the establishment of ozone, a gene that allowed our single-celled predecessors and every life-form that came after them to survive the sun.

When Earth's atmosphere started converting to oxygen, some scientists believe it caused mass extinctions. They've called it an oxygen holocaust, a revolution, a crisis. That's because the bacteria and archaea that didn't adapt to oxygen either died or were relegated to the nooks and crannies of the planet where oxygen couldn't reach. And they are still there, in our guts, in mud flats, in seeps and vents and cracks and crannies, hiding from oxygen. But for other types of bacteria and archaea, the oxygen holocaust was a bonanza, and the creatures that used oxidized minerals or evolved to use oxygen itself for survival would inherit the planet.

In other words, microbes created an environment, and nature selected those cells that could survive and thrive in that environment. Which means oxygen-respiring microbes are the ancestors of life-forms that need oxygen. You, me, my former rust-eating relative, indeed, all oxygen respirers, all plants, all animals, and all fungi are connected thanks to microbes.

CHAPTER

3

The Impossible Microbial Species Concept

I always sat in the same seat in class, in the middle of the third row—not in the front row, where I would be too obvious, but close enough to compensate for the fact that Professor Naeem is a soft talker and that decades of riding in clattering NYC subways have strained my hearing. I noticed everyone pretty much sat in the same seat throughout the semester, and I began to wonder, why did we sit where we did? And why always the same seat? Sometimes there was a bit of shifting around, especially in the first sessions of class, but very little. I asked one gal, who seemed annoyed when she found someone else in the seat she sat in the week before, why it bothered her, and she said, "I count on having one less thing to think about a day."

There are all kinds of studies that reveal students who sit in the front get better grades than those that sit in the rear, and student GPA even declines with every row, but whether that's because the students who get good grades are the ones that sit in the front or because the front helps them get good grades is up for grabs. Maybe there's a learning advantage to sitting in the front of the class, a combination of vision, hearing, attention, even eye contact with the teacher, which may increase a student's sense of personal responsibility to listen. I suppose after a night of beer pong it would be a lot easier to indulge in a quick nap if you are sitting in the back of the class rather than front and center.

Maybe seating is a measure of a student's enthusiasm for the class. My friend Jonathan is a biology tutor, and when he gets new students, he makes a classroom diagram and has them show where they "self-assort" in the class. "It helps me know how serious they are about learning," he said, "because they'll pretend they want to learn in order to fool their moms."

One afternoon I attended a lecture by the environmental geographer Ruth DeFries, whose work on satellite mapping of environmental impacts won her a MacArthur Fellowship. Most of the students up front seemed really engaged, but not the gal I sat next to in the back of the classroom. She spent the entire talk and its shocking slides of logged rainforests and desertification trawling through the romper offerings on the Juicy Couture website. I was a little surprised, but then, I kept forgetting that while for me biology class was an adventure, and the exposure to great thinkers a source of inspiration, for many of my fellow students, biology may well have been a required course they were just trying to complete with as little pain as possible, like filing taxes or getting your 10-year colonoscopy.

From my spot in the classroom, I earnestly tried to follow everything Professor Naeem said during those first few lectures about the origins of life, but I often missed sizable chunks of information as I struggled to write everything down. Further complicating my note taking was the rather depressing discovery that I could either use my reading glasses and see what I was writing or use my distance glasses and see the details of the slide. I settled for my distance glasses and wrote without looking at the page, struggling to keep up with Professor Naeem's lava flow of information about how species are determined. But when I got home and reviewed my notes, I couldn't read them at all.

I probably should have reviewed at least one of the student help sites on the internet on how to take notes in class. They all seemed to boil down to three simple rules, which I broke regularly: Don't take notes on everything the teacher says without actually listening; think about the topic (an impossibility if you are taking notes on everything the teacher says); and ask questions. I never got the hang of asking questions in class, mainly because I feared sounding foolish, so I just struggled along on my own, trying to catch up on class skills that I hadn't practiced in decades. Recently, I checked Dartmouth

College's advice page, which, among other tips, recommended students avoid writing notes on random pieces of paper.

I definitely would have benefitted from knowing this earlier. For years I have saved my son's old composition books in which he dutifully wrote three pages about the Trojan War or whatever before abandoning them. I was quite pleased to have a chance to repurpose many of them in my first semester. It was simply a matter of tearing out a few pages, including an abundance of doodles, some of which I saved because that's what mothers do. The chair desks in my lecture hall have narrow writing tablets about the size of a breadboard—too narrow to hold the whole composition book opened—so when I wrote on the left page, the right side of the notebook flopped over the edge of my desktop. During one lecture, while I was busy writing on the left-hand page, I noticed my notebook had captured the attention of the young man sitting next to me. I looked to see what he was staring at.

On the facing page of my composition book was a surprisingly accurate illustration of a naked couple screwing on a desk.

We both pretended we never saw the picture, but next class I noted he'd changed seats.

The first species concept that Professor Naeem explained to us was the biological species concept, which is basically about who has sex with whom. "There are problems with how we organize life," he began, "and people have tried to solve them lots of ways." The biological species concept determines species by whether they can produce fertile offspring. If they can't, they are different species. Humans can't mate with chimps and produce offspring at all, so we are different species. A donkey and a horse can produce offspring—a mule—but because it is infertile (though not celibate), which is as good as dead from nature's point of view, they're considered different species. But, he pointed out, because prokaryotes—that is, bacteria and archaea—don't reproduce using sex, this definition of species by fertility doesn't work for them at all.

Professor Naeem also explained the morphological species concept, which describes the relatedness of species based on body plan. This works fine for elephant seals and otters but not so well for prokaryotes since they can be hugely different on a genetic level even when they are similar in shape or size. Bacteria and archaea look almost identical but have less in common genetically than a rabbit and a lily pad. Then he explained the phylogenetic species concept, which says that species may be grouped by common ancestry. It works for humans, but it doesn't work for prokaryotes, because they can acquire genes from outside their ancestral line.

I was quite frustrated because none of the ways scientists define species applied to bacteria and archaea. But the truth is, categorizing life, while helpful when it comes to research, is a red herring. Life can be categorized by how organisms look, how they reproduce, how they eat, where they live, and where they get their energy, and which criteria one uses determines how everything is subsequently categorized. Ultimately, the organizing criteria we apply are subjective. We impose our biases on nature, and that's problematic because there is no perfect fit for all organisms. While we are keen to draw lines around organisms, nature has an annoying tendency not to color within the lines.

But so what if the criteria are subjective? I mean, what's the alternative? Aristotle launched the notion of taxonomic classification and binomial naming with his *History of Animals*. He grouped 500 types of animals by their similarities—things he could see: animals with blood and without blood, animals that live in water or on land, and so on. And he named them based on their "genus" (their group or race) and their "difference" within that group.

I asked my friend Neni Panourgia, a Greek anthropologist, about the origin of the word *genus*. Neni, a preternaturally cheerful woman, studies medicine and death and favors black leather pants. "To understand its meaning totally, think of this," she explained. "In Greece we have identification cards that contain genealogical information: your first name, your family name, place of birth and date, address, etc. For the married woman there used to be an extra field called genus, where the woman's paternal family name is mentioned. If I had changed my name to Gourgouris (her husband's last name), my ID card would read: first name: Neni, family name: Gourgouris, genus: Panourgia. What genus indicates is the biosocial origin—its 'difference,' the lineage that has created a singular individual as a part of a long line of individuals within the species." Using this example, Neni is one individual of the

Panourgia genus. Her living sister is another. And her unborn sister, had she married, would have been another. "However," said Neni, "I ate her." (Neni is a chimera, a single organism composed of cells from different fertilized eggs. She absorbed her twin in utero.)

The Aristotelian method didn't prevail in Europe, at least, not until the Renaissance. For most of the intervening centuries, plants and animals were classified based on their usefulness to us: warhorses versus workhorses, medicinal plants versus food plants. The bestiaries of Medieval Europe organized both real and imagined animals in rudimentary natural histories, usually with a moral lesson attached. For example, in *The Book of Beasts*, the unicorn, which is described as being as small as a baby goat and capable of being captured by using a virgin as bait, is characterized not as a ruminant but as a kind of proxy for Jesus, with hooves.

The Renaissance thinkers were inspired in part by the slew of interesting new species—like maize and pepper, llamas and possibly *Treponema pallidum* (the bacteria that causes syphilis)—brought home from overseas expeditions during the age of exploration. Unlike medieval scholars, they were less inclined to observe nature through the lens of the church. Inspired by classical Greek philosophy, they relied more on observation and inductive reasoning. (I'm guilty of being Eurocentric here. The Arab scientist Ibn al-Haytham, whose life bridged the 10th to 11th century, pretty much invented the scientific method. If it starts with *al*—algorithm, alkali, algebra—the concept is probably of Islamic origin.) But the classifications published by the European scientists that emerged during this time weren't coordinated. Everyone was doing his own naming, and centuries of unsystematic classification led to too many names. For example, by the early 1700s, the hippopotamus had accumulated five English names: river horse, sea horse, behemoth, river paard, and water elephant, not to mention numerous names in Africa's many languages.

That all improved later in the 18th century when the Swedish botanist Carl Linnaeus developed classifications based on shared similarities—kingdom, phylum, class, order, family, genus, species—and codified the naming practice of organisms with one generic name and one specific name. His is the system we still use today (although assorting organisms into the different classifications, and re-sorting the classifications of previous scientists, is an enduring preoccupation of evolutionary biologists). Linnaeus described the three king-doms as animal, vegetable, and mineral.* He did include microscopic organisms in his taxonomy as they had been seen in early microscopes and recognized as living things, though beyond that, they were a mystery. He lumped all of them into a group called Infusoria, at the bottom of which was *Chaos*, a hodge-podge of "microscopical somethings."

In the earliest days of microbiology, scientists classified microbes based on how they impacted human health. To this day, most people don't consider the microbe, not beyond the ability of a few species to make you sick. I've seen this in action in my biology book, like the chapter on viruses that mainly discussed HIV. Just to get it straight, microorganisms that cause disease are called pathogens, whether they are harmful to fungi, plants, animals, or other microbes. Obviously not all microorganisms are pathogens, but those were the ones early scientists were highly motivated to understand better.

The first attempts to classify microbial organisms came out of a need to comprehend the potential relationship between them and us, specifically their role in our diseases. And unsurprisingly, some of the first microbiologists were also physicians. Girolamo Fracastoro, aka Hieronymus Fracastorius, an Italian Renaissance man, developed a nascent theory of contagion in 1546, but he didn't realize that microbes were actually alive. That would have to wait until bacteria could be seen wriggling around by a 17th century Dutch draper named Antonie van Leeuwenhoek whose fascination with grinding lenses led to the invention of the first microscope, tiny as a watchmaker's tool. (I saw a reproduction in a glass case at Rutgers University, where I attended the fourth annual Bactoberfest, a beery potluck held for microbiologists in the lobby of Lipman Hall, Lipman being the professor who cultivated the soil bacteria *Azotobacter vinelandii* from a chicken nostril.) Leeuwenhoek, with his miniature microscope, took the first look at a monumental though miniscule new world, noting the different shapes of the microbes and their ways of moving about. The Italian Lazzaro Spallanzani, who was a rock star in the 18th century, discovered that microbes could be airborne and proved one of the basic presumptions of cell

---

*In a nod to his contribution, Linnaeus is considered the type specimen for *Homo* (genus) *sapiens* (species), although unlike other type specimens, Linnaeus isn't in a drawer in a natural history museum somewhere; he is buried pretty much right where you walk into the Uppsala Cathedral in Sweden.

theory, that all cells come from cells *and* that microbial cells reproduce by division. You can visit his bladder, in a virtual reliquary to science, in Pavia, his hometown.

My biology book pointed out that the study of microbiology has a rich history, "which was barely touched upon here." Nor did Professor Naeem delve too deeply into stories about the race to identify disease, which was, at times, reckless. Researchers who were studying pathogenic microbes in the 19th century didn't know exactly what they were dealing with, and the risks of infection were high. Unlike today, there weren't adequate prohibitions against using human test subjects, and some of those subjects died. As Paul de Kruif pointed out in his marvelous book, *Microbe Hunters*, it can be a "cruel road that leads to truth."

The 1800s was the era of the microbe hunters. First came elemental observations, that doctors' hands and surgeons' knives could transmit microbial infections, that living microbes rot meat and boiling the meat kills them. Louis Pasteur proved yeasts ferment grapes and barley—solving the 6,000-year-old mystery of beer—and they sour milk, which led to the eponymous practice of pasteurization, the heating of milk or wine to kill off those spoilers. He confirmed germ theory (*germ* started to mean "seed of disease" in English in the late 1700s), which says that disease is caused by microscopic organisms, and he discovered a little bit of disease, not enough to kill, could make us resistant to the pathogen, launching the notion of vaccines.*

To put this into historical perspective, at around the same time that doctors were figuring out they had to wash their hands between patients, the world's first oil refinery was built and the construction of the Suez Canal had begun. Technology was barreling ahead, but our knowledge of the incredibly impactful world of microbes was skin-deep. Microbes were mostly a mystery right up to Victorian times. After all, prior to Pasteur's germ theory, people thought angry gods, supernatural troublemakers, or imbalanced bodily fluids caused disease (and during the Medieval period, every healing was a miracle, which is why there was a saint for choking, a saint for rabies, and a saint for hookworm). Indeed, since the era of these discoveries until very recently, we

---

*Actually, the idea of vaccines has much older roots than Pasteur. Inoculation of patients against the smallpox virus by introducing bits of powdered pox scabs or fluid from pustules into superficial scratches was used in China and India as early as the 11th century, suggesting an implicit understanding of infection.

have believed the myopic story that microbes are all bad for us.

Over time, scientists discovered and described species, motivated, always during this period, by the desire to identify pathogens. The German physician Robert Koch proved the one-microbe, one-disease paradigm when he identified the bacterium that causes anthrax, which had been pandemic in Europe in the 1600s; tuberculosis, which had infected 70 to 90 percent of the urban populations of Europe and North America; and cholera, which killed hundreds of thousands of people in five pandemics during the 19th century. And Theobald Smith, one of the original four-member team that staffed the Bureau of Animal Industry (now the Agricultural Research Service), discovered that ticks could transmit disease, which opened the door to the discovery of other insect vectors, like the tsetse fly and sleeping sickness, the anopheles mosquito and malaria, and the *Aedes aegypti* mosquito and yellow fever.*

Throughout the 19th and early 20th centuries, scientists found and named bacteria, but there was no established nomenclature, and often scientists submitted different descriptive criteria for the same organism. Take *Candida albicans*, a single-celled fungus that is the source of yeast infections, mainly in the immunocompromised. Every time it was found on the human body (it's ubiquitous)—ears, gut, vagina, between our toes—it got a different name, which mucked up communication among scientists. Similar to the state of hippo taxonomy prior to Linnaeus, microbial taxonomy was a mess, and in many ways, it still is.

The biggest course correction in microbial taxonomy happened in 1977. Before then, there were thought to be two super groups from which everything else descended: cells with a nucleus (eukaryotes) and cells without a nucleus (prokaryotes). It's the most basic division among organisms. All single-celled organisms without a nucleus were considered bacteria. Carl Woese, an American microbiologist, looked at a particular ancient gene, the 16S ribosomal gene that codes for the construction of the ribosome, an elemental part of all cells. He found that based on differences in this gene, there were two types of prokaryotes—archaea and bacteria. He proposed we dump

---

*Being married to one of these early microbiologists was a perilous pursuit. The wives of most of the scientists I've mentioned were called assistants or secretaries, but more typically they were full collaborators in difficult and sometimes dangerous investigations.

the binary classification and divide life into a trinity of cellular life-forms—
eukaryotes, archaea, and bacteria—which meant classification of these
microbes really had to start over. That's a pretty big course reversal, and it
happened not that long ago, the same year that Jimmy Carter was sworn in as
president. I have T-shirts that old.

Is there that much difference between bacteria and archaea? Professor
Naeem pointed out the differences are profound; they have to do with cell wall
composition and how DNA is replicated and genes are expressed—in other
words, deep differences that are just about impossible to see and super hard to
comprehend. I imagined just about every kid in class was thinking: Does this
even matter?

Well, it does when it comes to classification because bacteria and archaea
account for the majority of diversity among species. Archaea are easier to grasp
than bacteria, but not because they are less diverse. Rather, there's less known
about them, so their simpler classification scheme is ultimately misleading, and
new phyla of Archaea are constantly being proposed. Those identified seem to
be mostly extremophiles, and our textbook designated two phyla. One was
Crenarchaeota (from the Greek for "ancient fountain"), mostly microbes that
can live in incredibly hot and/or acidic environments, like marine and terrestrial volcanoes, an adaptation that might have its origins in the beginning of
terrestrial time. The other was Euryarchaeota (from the Greek for "wide," as in
broad, and "ancient"), mostly either halophiles, which live in extremely salty
habitats, or methanogens, which live in oxygen-free environments like cows'
stomachs and produce methane as a by-product.

But bacteria? Yikes. "There are lots of groups," warned Professor Naeem.
"Our book just looks at a few. But you still have to learn them." I tried to get a
grip on an overview, at least some minimum classification for them, but everywhere I looked I found different compilations, and they all approached bacterial classification using indefinite pronouns: "*Some* representatives of various
phyla of Bacteria" and "*Some* major clades." Then they listed a handful of iconic
groups like Actinomycetales, which are common in soil and in dental plaque
and make antibiotics we use. Chemoautotrophs get their energy from sources
like hydrogen sulfide, ammonia, and methane, and they play an important role

in the nitrogen cycle of the planet. Thermophiles live in improbably hot habitats and are likely among the most ancient bacteria. Cyanobacteria are photosynthesizers that produce oxygen as a by-product, and the Enterobacteriaceae include inhabitants of animal and human intestines, like *E. coli*. Pseudomonads are common soil bacteria and the source of many plant pathogens, and Spirochaetes are spiral-shaped bacteria that whirl through gooey substances like mud and mucus, the same critters that brought us Lyme disease and syphilis. But every list was qualified with versions of "we have only scraped the surface of bacterial diversity."

After a really unfair amount of confusion, I began to see that what we were being taught was ultimately a sampling of the unknowable. Identifying bacterial species is complicated. They are incredibly diverse, and it is very hard to identify them by their genome because the genes that are present in a bacterium don't tell you which genes are expressed and which genes are just silent and conserved over the eons. I mean, we share 99 percent of the same DNA as chimps, but it is pretty obvious we are not expressing the same genes, and we are not the same species. As the biologist Lynn Margulis wrote in *Symbiotic Planet*, "Even if there were identical gene sequences in the bark of a banana tree and the skin of a dog, we would still classify a dog not with a banana but with wolves and jackals." What distinguish us are not only which genes we have, because there are many we share with our evolutionary ancestors, but also which genes are turned on and which genes are turned off.

The second obstacle to bacterial identification (and hence classification) is that they can be so particular to their environmental niche that they mostly can't be taken outside those environments and grown in the standard petri dish of nutrients to be studied. The vast majority of the microbes you can see through a microscope are hard to cultivate in the lab. For example, anaerobic bacteria die at the first whiff of oxygen; certain heat-loving bacteria freeze to death if temperatures drop below 100°F. Or the environmental niche a bacterium needs to survive depends on the presence of other bacteria. In some cases, a quorum of bacteria is necessary to produce enough of a by-product, like CO_2, to sustain a community of CO_2-dependent bacteria. Even more, there are bacteria that live at the interface of two bacterial by-products, where a gas diffusing toward them from one type of bacteria meets the gas diffusing toward them from another type of bacteria. That almost implausibly scant point where the two gases meet is their ecological niche—not so easy to reproduce in a petri dish. Bacteria and archaea that can't or haven't been cultivated can be named, but they have a prefix, *Candidatus*, meaning they exist in a kind of taxonomic limbo; they're candidate species—species in waiting.

To further complicate the cultivation problem, some bacteria have evolved to count on the genes of other bacteria to survive at all. In 2015, biologists found cells with genomes so small they are on the edge of the genetic requirement for life, and so they are probably dependent on other life-forms for some of their functions. To cultivate these, you'd have to grow them out "with their friends," said Philip Strandwitz, a microbiologist at Northeastern University in Kim Lewis's Antimicrobial Discovery Center, which looks for new antibiotics using co-culturing techniques. But the problem of classification remains. "At what point do you decide what is a single species composed of more than one collaborating organism and what is two species living in a mutualistic relationship?" asked Professor Naeem. One thing is for sure: Of the 1 percent of bacterial cells that have been successfully cultivated in the lab, most are pathogens of humans, animals, and plants. Interestingly, the bacteria that make us sick are the ones most willing to oblige us in the lab.

And finally, bacteria are difficult to classify because they evolve so quickly; due to the effects of random mutation and horizontal transfer of useful genes, by the time the scientist has completed her research and published a paper describing the bacterium, it has evolved into something else. Most bacteria have never been studied in part because they mutate so quickly. "Bacteria can evolve, spread genes, and adapt so fast," wrote Lynn Margulis, "that no possible human taxonomy can truly contain them."

Under these problematic circumstances, similarity of their DNA is the preferred approach for classifying prokaryotic species. Since the inauguration of Carl Woese's three-domain system, bacterial trees of life have been based on the 16S ribosomal gene. Each bacterial species' 16S gene is unique. If two bacteria share 97 percent or more of the 16S ribosomal gene, they are considered the same species. This has given rise to a new classification term: the phylotype, aka the genomic species. The cost and ease of genetic analysis has led to a frenzy of testing for bacterial genes in everything from a cheek swab to a teaspoon of soil, and this has revealed a vast diversity of bacteria—or bacterial genes anyway—with estimates of maybe a trillion species. But those groups contain so many diverse types that if you applied the same criteria to animals, humans and worms would be the same species. It seems there are as many kinds of bacteria as there are bacteria.

I was so utterly bugged out by the fact that bacteria weren't falling into neat categories I decided to reach out to Professor Naeem for some guidance as to how to wrap my mind around the, well, uncooperative nature of bacterial species. I knew Professor Naeem had weekly office hours, but for some reason I didn't really understand they were held for the purpose of explaining material. In a way, I felt too far behind, and I didn't want to reveal my ignorance. But I made an appointment with him anyway. I learned many things at that meeting, including this: If you ask a scientist a vexing question, don't be surprised if you get a philosophic answer.

"So what about a microbial species concept?" I asked.

"I think you should read Michel Foucault," he said. "He had some pretty interesting ideas about classification."

In the preface of Foucault's *The Order of Things* he describes a short story by the Argentine fiction writer Jorge Luis Borges called "The Analytical Language of John Wilkins," which cites an ancient Chinese encyclopedia entitled *Celestial Empire of Benevolent Knowledge* that divided animals into those (a) belonging to the emperor, (b) embalmed, (c) tame, (d) suckling pigs, (e) sirens, (f) fabulous, (g) stray dogs, (h) included in the present classification, (i) frenzied, (j) innumerable, (k) drawn with a very fine camel-hair brush, (l) et cetera, (m) having just broken the water pitcher, (n) that from a long way off look like flies.

Bear in mind that Borges was a surrealist, probably not the best place to start on any quest for the order of things. Indeed, it's a totally silly list, a fact some folks have missed, but I think they can be forgiven considering how hard it is to penetrate Foucault's brilliant, albeit airless, writing, a style which struck me as so old-fashioned I was surprised to learn he was the same age as my father (Foucault died of AIDS in 1984). Borges's bogus classifications got Foucault thinking about the capriciousness and cultural bias of the whole exercise, and he argues that historians and scientists are incurably influenced by their cultural, institutional, and religious contexts.

This was Professor Naeem's point, too. When it comes to classifying nature, not only are our prejudices evident from a historical perspective, but we are prejudiced every day by the fact that our understanding of life is locked into a scale relative to our experience of the world. We need to classify organisms in order to study them, and so it will happen, but every biologist I talked to about this seemed to think the whole concept of species just falls apart on a microbial scale. "The Linnean fantasy of a divine order throughout nature that included unambiguous species," wrote Nicholas Money in the journal *Fungal Biology*, "rests on an unstable philosophical foundation."

Maybe the notion of species itself is a fallacy. "I would not hesitate to say there are no species in nature," the mycologist Tom Volk told me. "Look, the main problem with species concepts in general is it assumes evolution is done," he said. "And it's not." It seems the evolution of prokaryotes happens faster than our definition of species can handle. It's practically surreal.

I often peeked at the other students' note-taking habits in class. Some used their laptop computers, but most wrote in spiral notebooks, and from what I could see, everyone but me was doodling like crazy. (Of course, there are garden-variety doodlers and there are great doodlers. A Picasso doodle will set you back $6,500.) At first, I thought this indicated that, comparatively speaking, I might do alright since I was taking notes so diligently. But then I read a study published in the journal *Applied Cognitive Psychology* in 2009 that suggested doodlers remember more than nondoodlers.

The author speculated that when you doodle, you don't daydream. Daydreaming actually requires a certain amount of mental agency. Let's say you start thinking about your weekly menu plan and what you need to get at the grocery store on the way home. Maybe you should pick up the chicken now and freeze it for later instead of buying it in 2 days? That process, of prioritizing and decision-making, requires executive functioning, an umbrella term for the management of cognitive operations, and executive functioning uses mental resources that displace the energy you need to create new memories.

In contrast, doodling is the opposite of executive functioning. Making a doodle is pretty undirected. As a result, it doesn't interfere with your ability to remember. Nor does pen twirling (a how-to YouTube video on pen spinning has over 1.3 million views, and there is such a thing as competitive pen spinning—the sport is particularly popular in Japan, China, and South Korea) or other fidgety behavior, which some folks who suffer from ADD say helps them stay focused. The problem, of course, is the focus of the person sitting next to an impressive pen spinner. It looks like baton twirling and is kind of mesmerizing. I spent a good hour being distracted by videos of the Wiper, the Shadow, the Charge (that's what drummers do with their drumsticks), and the Sonic, where the pen is flipped between fingers faster than the eye can follow. I almost had a fender bender on Fordham Road in the Bronx because I went into a virtual trance staring at the superlative pen twirling of a passenger in the car next to me.

After a few weeks, Professor Naeem began posting his slides on Course-Works, an online site that offers details about the class. This was a relief as it took some of the pressure off my note-taking skills. CourseWorks or something like it is common in college today. After logging on and finding my course, I had a choice of pull-down menus. I found the lecture slides under Files and Resources, but there were other menus, like Syllabus. I initially didn't bother opening it because I thought I knew the reading assignments until one night I was about 5 grueling hours into a dense chapter about the sensory systems of plants—which had nothing to do with single-celled life-forms—and realized, okay, this is the WRONG DAMN CHAPTER. I thereafter checked Syllabus religiously and in that fidgety way that you check your watch and ticket and carry-on luggage every 5 minutes while you wait to board your airplane.

Unfortunately, my notes were not particularly helpful when I compared them to the slide show, which was stuffed with detailed illustrations of gram-negative and gram-positive bacterial cell walls, one of the ways microbiologists identify them, and the parts of the cell that read to me like Greek: peptidoglycan, cell wall polymers, LMW SLP, CwpV. . . . It wasn't long before I was on Khanacademy.org looking for explanations and fantasizing about the faceless narrator Saul, whose calm repetition reminded me of Bob Ross, the "happy little clouds" PBS painter. Ross is well-known among autonomous sensory

meridian response (ASMR) junkies—ASMR is also known as a brain orgasm—
who get tingly feelings of relaxation from triggers like listening to a soft voice
or repetitive tasks like turning the pages of a book. Weirdly, Saul's videos did
the same for me.

I ended up devising a note-taking strategy that I thought was pretty smart.
I started numbering the slides in my notebook as they came up during the lecture so I could coordinate them with the slides Professor Naeem posted afterward. And to help read all those incomprehensive technical abbreviations on
the lecture slides and ensure my notes were readable, I went to the eyeglass
store and got a pair of bifocals, my first.

A few weeks into class, I received an email about the Ecology, Evolution
and Environmental Biology department's annual fall BBQ. I didn't realize E3B,
as it's called, *was* a department, and a new one, too, having been established in
2001. I asked my advisor, Matt, what lay in store for these E3B students I saw
napping with their coats over their faces in the lounge or lugging backpacks
like sacks of stones. He told me some graduates go on to medical school, some
into graduate or PhD programs, others directly into research. And, this being
New York, a few end up in finance.

The BBQ was held in a little piazza on campus. Potted privet hedges defined
a perimeter space that was outfitted with round tables and chairs, and at the far
end, a buffet table serving BBQ ribs and chicken, corn bread, and greens, and a
bar stocked with art-opening-grade wine and warm beer. I didn't know anyone,
of course, and so I took a place in the buffet line, usually a decent way to meet
people. Though I fished with hopeful eyes the pierced gal in front of me and the
shy bald man behind me, I didn't get any bites, and so after a bit of lame banter
with the catering guy scooping ladles of corn salad, I filled up my tray and
turned to face the tables.

There were about 25 people seated. It looked like the students were all sitting at tables on one side of the piazza, most my own children's ages I assumed,
as I only saw the tops of their heads and their thumbs pecking out text messages
on their phones. The faculty seemed to be seated at the other end of the piazza.
They were the biologists, I guessed, and around my age—the women wearing
comfortable tunics and handmade sweaters and dangly earrings, their hair
unstyled, lovely and unfussy, bohemians with gigantic vocabularies. The men

were equally unpretentious. Wearing their field clothes, they looked like they'd just come from a nesting crane stakeout and had scrounged around in the back of the jeep for a jacket to wear to dinner.

I stood, tray in hand, looking to the left and to the right. I had no idea which group was my species. I didn't know where I belonged.

It occurred to me the reason why I was so obsessed with finding some kind of order in the microbial world was because I felt so out of place in school. I was terribly self-conscious, like a visitor from another country, or another era. I felt my awkwardness most when classes let out. All the students poured out onto the stone and brick walkways, heading to their next class, a mass of sneakers and acne and cell phones on the move. I navigated through them, staying to the right of the path and observing with the immaculate manners of a good tourist.

That feeling was exacerbated in the close quarters of recitation, where I shared a classroom and conversation with maybe a dozen students. A young grad student named Kaiya conducted my recitation section. She was bright and brown-eyed and quick and abrupt in her movements, like a bird, which, coincidentally, she studies. Attending recitation can bump up your grade a bit, not only because you get some reinforcement of the study material, but also because you get a kind of advantage for showing up. Professor Naeem had explained that if your grade is borderline, a good record of attendance to recitation can weigh in your favor. The way the TAs keep track of attendance is a sign-up sheet, enlivened by a personal question. The first day it was "How many biology courses have you taken?" I was the last person to sign up, so I got a good look at my fellow classmates' backgrounds. Most had two, some three, courses under their collegiate belts. I wrote down my number: zero. Another day the question was "What is your favorite organism?" A lot of people wrote down cats. And another, when we studied the various nervous systems, "What is your favorite sense?" Feeling a little punchy, I wrote down ESP.

Recitation was held during the lunch hour and consisted of some review, but mainly it was an opportunity to get your questions answered. In general, I was too bewildered by the material to even form a question. What was the difference between meiosis and mitosis? I'd lean forward in my chair, trying to follow as Kaiya scribbled diagrams on the board—not a chalkboard like in my day, but a white plastic one—using colored markers that squeaked as she drew diagrams of sausagey chromosomes linking up and doubling and then more drawings showing their limbs crossed over like gingerbread men who had merged arms in the baking tray. As she drew, her diagrams tilted to the right until they crowded in a puddle at the bottom of the board. And then she'd suddenly stand back and say, "Clear?"

Everyone mumbled yes, and I did, too, even though it was not, and someone asked if this was going to be on the test.

I got a pretty good look at my fellow students in recitation. There was a smart girl with an asymmetric hairdo who asked precise questions, and a beautiful Sicilian girl with bitten-down nails whose questions smoothly incorporated terms like *protein kinase*. In her Italian accent, it sounded like she was ordering something delicious from a dinner menu. I was particularly interested in a tall young man with a head of black curls who was very candid about what he didn't know, and I admired his gumption to ensure he got his answer. "Well, I'm not totally clear," he'd say. I came to count on him for my own clarifications.

Kaiya encouraged us to bring our lunch. I brought sandwiches with homemade bread and grilled steak from the night before and pieces of soft quiche, while my fellow students munched on chips or takeout. There are a number of Chinese food carts parked right outside the campus entrance gates. They show up at lunchtime like street-corner umbrella salesmen when it starts to rain, and within minutes, long lines of students form to place their orders.

Before recitation one day, I got in the longest food-cart line, what I thought was a sure sign it would have the best food. I also got on my cell phone like the other students, though I don't use it for Twitter or Instagram. I just read the day's pitches from the variety of nonprofits whose mailing lists I am on and from which I am too guilty about my own luck in life to unsubscribe. The dingy cart was colorfully illustrated with photos of shiny dumplings, shiny bok choy, shiny noodles. As the students peeled off with their Styrofoam containers and I came closer to the cart, I could feel its heat, like a lamp that had been left on too long, reeking faintly of soy sauce and gasoline. I chose similarly to the other students: two (from a choice of six) items for $6. The vendor peered down through the window of his cart, a foot higher than my head, sticky steam billowing around him. Stir-fried tofu and cabbage, please. He grabbed a large Styrofoam container and slapped huge servings of tofu in its chili sauce and steamed cabbage into their separate compartments and dumped a mound of white rice on top. He closed the container and passed it through his window, a string of limp white cabbage hanging from the side.

During recitation, I inspected the steaming food. It filled the room with a strong hot smell, like the food cart on a plane, and I looked around self-consciously. While I listened to Kaiya explain the difference between systematics (I am descended from apes) and cladistics (apes and I share an opposable thumb), I sampled my lunch. The food was tasteless and overcooked; one particularly chewy morsel turned out to be a rubber band. Nonetheless, I ate until I was full, and still there was enough food left over to feed me two more times, answering the riddle of the long line at that particular cart.

I had a running fantasy about bringing lunch for my fellow recitation students: containers of chicken stew with biscuits, sandwiches with mortadella and pistachio pesto. But I never did. There was something show-offy about this impulse even though I imagine they would have happily gobbled down anything I offered. At one point in recitation, a young woman named Catalina wondered what brought me back to school. I was so happy to be asked. I explained my interest in microbes, that I was an author who'd written about fungi, and then the tall student whose questions I had come to count on said he'd heard a talk about mushrooms once, and was interested. I recommended he join the New York Mycological Society, where he could learn all about them. As I chattered away happily, suddenly existing as who I am, the result of my maturity and knowledge, not some ghost of a person on the periphery of the college experience, he looked me up on his cell phone and asked if I was Eugenia Bone. I puffed up a little thinking, well, they'll find out I am rather accomplished, and then he held his phone up for me to look at. On the screen was a piece of ricotta pie from one of my recipes.

"Is this you?" he asked.

Well, I suppose if I was classified based on recipes, yes, I am a piece of ricotta pie. That young man could search my name differently and I would come up as an author. And in another few years, as my online presence changed, maybe as a jar of jam.

That's when I realized nothing can really be classified with total accuracy and in perpetuity, because everything—me, a bacterium—can be understood in multiple ways that are always changing. A beneficial mutation on the genome of one bacterium can spread across a bacterial community, and the chances of even more sharing increases each time a cell with that mutation divides. In an increasingly connected world, that mutation can be shared among individuals in an ever-larger population. And the larger the population, wrote the microbiologist Stephen Giovannoni in *Microbes and Evolution*, the more efficient its evolution, because there is more opportunity for the mutation to spread. The immense bacterial domain with all its zillions of constantly changing cells is like one gigantic gene pool, one "worldwide superorganism," according to Lynn Margulis. That's why we can't cultivate them easily, and that's why we can't classify them easily.

Because they are all connected to each other.

CHAPTER

4

A Marriage of Microbes

T he worst thing about being a student is taking tests. We had four during the course of my semester. I was worried about them for three reasons. Number one: Anxiety. I hadn't taken a test besides the kind of pop stuff you see in women's magazines for years. ("Quiz alert! What does your ponytail say about you?") Two: Impulsiveness. When I was taking the online biology course, I had to constantly fight the impulse to take the test before I'd finished the reading material, which could qualify as a kind of personality disorder, something like when you suddenly feel the desire to swerve your car off a bridge. Three: Humiliation. I didn't know if I could cope with a poor showing. I am used to doing things well, in part by outsourcing the things I do badly, like ironing, computer problems, and haircutting. I know who I am, and I am not someone who knows how to navigate Apple TV. I think we direct our lives toward chores and goals for which we have a natural ability or passion, but I also think we steer away from things that are particularly difficult to master or loathsome to learn, from things we've failed at before.

The most profound challenge I faced by going back to school was trying to be good at something that long ago I determined I sucked at: science, math, and memorization. I've been avoiding them for decades. I've performed a talk about my adventures picking mushrooms dozens of times, and I still can't get up in front of an audience without note cards. (Glossophobia, the fear of speaking in public, is the #1 phobia of Americans. Athazagoraphobia is the fear of being forgotten or forgetting. Obviously, I suffer from them both—also gelotophobia, fear of being laughed at, no doubt because of my athazagoraphobia. I think I can also add hellenologophobia, fear of complex scientific terminology, and kakorrhaphiophobia, fear of failure.) And math? My checkbook balancing technique is a baroque system so tied to the statement's layout of debits and credits that when I changed banks and was presented with a slightly different accounting system, it took me 4 months to figure out how to balance again—and not without a few irate and, in retrospect, unwarranted telephone calls questioning the bank's integrity. But once I was back in school, I had to face those gremlins again.

Not that I didn't look into alternative options. There are many sites on the internet that explain how to study for a test. Business Insider's "10 Study Hacks That Will Help You Ace Your Final Exams" includes rewarding yourself with treats, recommending gummy bears versus "shots of liquor . . . the idea is to retain information, not black out," cramming effectively by reading upside down and aloud, and teaching a class on the subject to stuffed animals. I tried to console myself, saying, hey, it's not like I am taking the bar (about 50 percent flunk) or the CPA exam (about 50 percent flunk) or even the 20-minute driving test. Actually, the driving test is not an encouraging analogy, as I am one of the approximately 50 percent that flunk it the first time.

I found our first study sheet on CourseWorks, with a cheery note from Professor Naeem.

Dear Environmental Ecologists: Attached is review sheet one. Our first week was devoted to the origin of the universe within which life emerged. We considered how the cosmos essentially consists of small numbers of parts that are assembled in a large variety of ways to produce diverse arrays of entities. Six quarks, six leptons, and five bosons make all the elements in the universe. While there are only 118 elements, it does seem, if we had enough energy, we could keep adding hadrons and electrons till we got more and more massive atoms, though it seems that the heavier they get, the less stable they are. It takes the energy of supernovas to make the heavy elements, but once we have them, we can make a near infinite number of molecules. . . .

And so it went, in Naeem style, a whoosh across the eons, from the big bang to multicellular creatures, for another 1,000 words or so, concluding with a promise that everything we needed to know was on the study sheets. "That's my deal with you," he wrote. While I appreciated knowing what I was supposed to know, I was nonetheless completely freaked out by the sheer quantity of material in the study sheets—over 150 terms ("hopefully, you've encountered them all before") and three pages of questions like "When we look at main sequence stars ranging from 60,000° to 2,500°K and consider temperature, mass, size, luminosity, and longevity, how does our Sun compare to most stars?" and "What role does the R group on the amino acid play in protein structure and function?" It was mind-boggling.

These five single-spaced pages covered five chapters in our biology class and two lectures. I bought a mechanical pencil with the thinnest lead I could find, and in tiny mouse script that would have done the tailor from Gloucester-shire proud,* I filled in the definitions for terms like *plasmid, pseudopeptido-glycan, murein* and wrote the answers to questions along the margins of the paper, little arrows tracing paths to answers like "pilus connects 2 cells, plasmid replicates and rolls out ccs like tape, replicated in recipient cell." I wrote in such tiny handwriting and made such miniscule diagrams that in order to review my study sheets, I had to use my husband's more powerful reading glasses.

Early on I had spotted a student who was a bit closer to my age, and we nodded to each other as we found our seats in class (she sat in the same row as I, all the way to the left directly in front of the podium) in unspoken recognition of our similar situations. A graphic artist with tongue piercings—an indicator that she was at least 15 years younger—she stopped me in class shortly after we had received our review sheets. "What do you think of them?" she asked. I told her they were overwhelming. How was she coping? "Stress soup," she said, widening her eyes for emphasis. "I am spending *way* too much time on this course!" So was I.

It took me hours to read a chapter. I couldn't imagine how the other students handled a full course load, because just this one class was turning me into a jerk. When my husband asked if I could find him a copy of our insurance documents, I looked up from my books and snapped, "How can you not know where we keep them after 25 years of marriage?" I stopped returning emails,

*The mice that finished a tailoring job in Beatrix Potter's favorite tale left a buttonhole incomplete and a note in script relative to their paw size that said "No more twist!"

or responded minimally. I began to use "K," a sour, reluctant way of saying yes. I gave up the gym and made meals of Ritz Crackers and peanut butter and started sneaking cigarettes. It felt like I was 18 again, only without the waistline.

When I was young, I learned that bacteria, cells without a nucleus, evolved into eukaryotes, cells with a nucleus, and then those cells evolved into primitive mollusks and insects and fishlike things on the one hand and plantlike things on the other, and then these things eventually crawled out of the water and spread across the land and evolved some more, until we ended up with, I don't know, Dolly Parton. That was enough understanding of evolutionary biology for me to have a career developing recipes. And it's kind of true, but when I went back to school, I learned I had missed something really fundamental and important. Multicellular life started when two individual cells started working together. One cell didn't just evolve into something more complex. Complexity is the result of cooperation between two cells. If life can be summed up in an equation, it would be, to quote the biologist John Archibald, "One plus one equals one."

It turns out we are connected to bacteria in the most intimate way imaginable. Take a deep breath. Breathing is one of the few systems in the body that is controlled unconsciously and consciously (well, that's not totally true, because even though you can hold your breath until you pass out, once you are unconscious, you start breathing again). Breathe in and the oxygen is picked up by red blood cells that distribute the gas to a tiny bean-shaped mini-organ in every one of our cells called the mitochondrion. Mitochondria are like furnaces. They burn the oxygen and use the energy to convert glucose, a sugar molecule derived from carbohydrates that you picked up from eating dinner, into fuel to run, basically, you. Breathe out and carbon dioxide, the by-product of burning that oxygen, is exhaled. Those little organs, the mitochondria, produce 90 percent of the energy needed to sustain life and support the function of the big organs, like our brain, heart, and lungs. Mitochondria are exceedingly important. Without them, we'd be pretty lethargic, to say the least. Actually, mitochondrial disease is no joke. Because they are part of our most essential machinery, dysfunctional mitochondria can affect the heart, brain, muscles, and lungs, causing seizures, strokes, severe developmental delays, inability to walk, talk, see, and digest food, among a host of other problems. But here's where it gets surprising. They aren't animal in origin. They are ancient bacteria.

Mitochondria are fundamental to our ability to be more complex than bacteria, and yet they descend from a free-living bacterium that lived in the ocean 1.5 billion years ago. A host cell, probably a type of archaea, swallowed but did not digest a bacterium and utilized that bacterium for its energy-making ability. Obviously, the details of this encounter are a little fuzzy. Maybe the bacterium was attacking the host cell but became domesticated. Or maybe the bacterium produced something the host needed and so was enslaved. Anyway, every time the host cell divided, the bacterium inside divided as well, distributing its progeny into the host's progeny. Over time, lots of time, the bacterium lost the genes necessary to survive as a free-living organism and handed over many of its functions to the host cell, thereby becoming an organelle. A mitochondrion is a genetically whittled-down version of its ancestor. It's no longer free, but it is still producing the energy that it once produced for itself alone. In the secret privacy of our cells resides the descendant of another creature altogether.

Professor Naeem explained this to us in class, and it took everyone a few minutes to digest the implications. We are classified as animals, but actually, we are also bacteria. Not just the host for bacteria that we acquire through life on our skin and in our guts, but we have ancient bacteria incorporated into our cells that we depend on, and we pass those bacterial cells through the matrilineal line, from mother to daughter. All creatures who are eukaryotes—that is, everything but bacteria and archaea—are made of cells that contain mitochondria. This billion-plus-year cooperation between cells is called the endosymbiotic theory (*endo* from the Greek for "within," and *symbiosis*, "living together"), and it didn't go mainstream until 1981—the same year MTV went on-air with 24-hour videos—with the publication of *Symbiosis in Cell Evolution* by the biologist Lynn Margulis. She had been working on the concept since her 1967 paper, "On the Origin of Mitosing Cells," which offered physiological, biochemical, and paleontological evidence for the hypothesis that the mitochondria in our cells, in all cells with a nucleus, are ancient bacteria. Margulis was

Lynn Sagan then, married to the astronomer Carl "billions and billions of stars" Sagan, whom she met at the University of Chicago. (As one friend put it, "Imagine that dinner table conversation!") Margulis was precocious, bold, and brilliant. She started college when she was 15 and married Sagan after she got her BA at the tender age of 19.*

The endosymbiotic theory was initially dismissed. In the early 1970s, two biologists at the University of Indiana refuted the concept with their paper, "The Non symbiotic Origin of Mitochondria," and the idea had been labeled a bad penny that keeps turning up. The notion that mitochondria may have bacterial origins had been batted around for 50 years. As early as 1918, the French biologist Paul Portier tried to culture mitochondria in the lab, believing that not only were they bacteria but they were capable of living independent of the host cell (they are bacteria, but they can't live independently). But then the evidence started to pour in.

The mitochondrion is distinct from other tiny organs in the cell. It resembles a bacterium. It has its own membrane and contains its own DNA. Mitochondrial DNA is not in a nucleus like the host cell's DNA; it is formed in a ring of genes that looks like the circular plasmids common to bacteria. Mitochondria have their own ribosomes, which manufacture proteins that are more like bacterial ribosomes than the host ribosomes elsewhere in the cell. In fact, you can destroy both mitochondrial ribosomes and the ribosomes of free-living oxygen-respiring bacteria with the same kind of antibiotic (obviously this is not an antibiotic we use medically). And the reason that antibiotics don't hurt you is because your eukaryotic cell is more closely related to archaea, the original host cell, than to the mitochondrion it houses. If you are injured and your cells break open and release mitochondria into your bloodstream, your immune system will attack them, mistaking these 2-billion-year-old symbionts as the bacteria they originally were.

There's more. Mitochondria multiply by doubling up on their genetic material and then dividing in two, with each daughter getting the same set of genes, just like bacteria, and they do this independently of the host cell. But

*Another paper came out of Norway the same year that described a similar endosymbiotic hypothesis. Few remember Jostein Goksoyr, though he does have a species of bacteria named after him: *Bythopirellula goksoyri*.

what governs the multiplication of mitochondria—like, why divide now?—is often dictated by the host. For example, when you work out and your muscles need more energy, your muscle cells tell their mitochondria to increase energy production. That's achieved when the mitochondria increase in numbers by dividing. One mitochondrion becomes two, the two become four, and so on, until the cell's energy needs are met. You can have multiple mitochondria in cells that use a lot of energy. Additionally, when we have children, Mom and Dad's genes recombine to create a new mix in our kids, but our children inherit only Mom's mitochondrial DNA, not Dad's—the genes in the mitochondria are passed to the next generation in Mom's egg cells, like an heirloom. Inherited mitochondrial genes are distinct from the genes in your child's nucleus, which is why as long as you've got some mitochondrial genes to look at, researchers can trace the genetic lineage of mitochondria back in time. Some researchers have reported they've traced matrilineal mitochondria back 200,000 years, to "Mitochondrial Eve," a common ancestor for humankind from East Africa.

There is no Mitochondrial Adam, by the way. Mitochondria provide the energy sperm need to get where they are going but don't make it into the embryo; why and how they don't is unknown.

In fact, scientists don't know why endosymbiosis happened at all, though not for lack of trying. It doesn't help that if the host cell was archaeal, they don't know what kind of archaeal cell it was, nor how it made its living. But let's say the archaeal host couldn't respire oxygen. In a rapidly oxygenating world, it would be beneficial to control an oxygen-respiring bacterium to do the job. There's evidence for this. Mitochondrial DNA looks a lot like the DNA of bacteria from the order Rickettsiales, very tiny oxygen-respiring predators that cause diseases like typhus and Rocky Mountain spotted fever. It is possible a bacterium like Rickettsiales invaded the host cell only to end up staying put and over time adapting to its new digs by assimilating and becoming less virulent. Another hypothesis suggests the host cell was similar to the methanogenic archaea that live in cows' guts, which might have depended on hydrogen the way our cells depend on oxygen. In this scenario, the archaea engulfed a bacterium that produced hydrogen. From the host's point of view, it would be a good deal to control an energy source, sort of like buying a house with solar panels.

Over time, the bacterium would have adapted to environmental pressures and respired oxygen. There is evidence for this, too.

One of the biggest questions is whether or not the host cell had a nucleus when it engulfed the bacterium, or whether the nucleus evolved later, after the capture. "The nucleus is a huge, huge problem," said the microbiologist Moselio Schaechter. "There are many opinions as to how the nucleus formed and no evidence." My biology book explained the standard spiel. An ancient relative of modern-day archaea enfolded its membrane, kind of the way a paper cup folds in on itself when you crush it around the mouth with both hands. Eventually the folds and the spaces between them were enclosed by a reformed cell wall, with the DNA trapped in the center. When you look at a cross section of a eukaryotic cell, you can see what looks like an enfolded membrane captured within the cell and surrounding the nucleus and its mass of DNA. But there are other hypotheses. For example, one suggests the nucleus formed its membrane to protect the host's DNA from bacterial attacks.

However we got it, a nucleus allows the eukaryotic cell to pack in more DNA. And in the case of human beings, we have *lots* of DNA, like 6 feet of the double helix rolled and compressed and packed into a compartment 100,000 times smaller than a grain of sugar. And our nuclei are hardly the ultimate hoarders. An onion packs 12 times as much DNA in the nucleus of its cells, and an amoeba, which looks like a loogie someone spit on the sidewalk, only about the size of a speck of pollen, has 200 times more DNA than, oh, Stephen Hawking.

Of all that, less than 2 percent of our DNA is used to make proteins; a larger portion regulates that 2 percent by determining which genes are turned on and which are off, and the rest don't code for proteins. These are often called junk genes; they may have been useful in our evolutionary past and may be useful again. I like to think of these genes as junk you keep in storage. You got the skis because you rented a house near Hunter Mountain for a few years back in the '90s. You keep your skis because, well, maybe you'll retire to Colorado. Or as the biochemist Nick Lane observed in *The Vital Question*, "Genomes do not predict the future but recall the past." Each one of our cells contains all of our genes. What differentiates, say, a liver cell from a neural cell is a matter of which genes are turned off and which are turned on.

Having a nucleus is the defining attribute of the eukaryotic cell. However, when the mitochondrion took over energy-producing duties, it freed the host cell from the nuisance of making its own food, and as a result, all kinds of evolutionary opportunities arose, include multicellularity. It's analogous to the advantages of an agricultural society versus hunter-gatherers. If enough food can be produced using less of a community's energy, then there is energy leftover for non-food-driven occupations like science and art and skiing to flourish. As a result, the eukaryotic cell is 10 times bigger than a bacterial cell, and it's way more tricked out. Endosymbiosis was as big an evolutionary leap as the beginning of life itself. It's the primal arrangement that permitted the evolution of the cell type that makes us.

And there wasn't just one endosymbiotic event. There was another.

Maybe a hundred million years after mitochondria were established in eukaryotic cells, a eukaryotic cell with a mitochondrion engulfed a cyanobacterium—the kind of photosynthetic bacteria that oxygenated the atmosphere—and over time, a lot of time, that cyanobacterium evolved into a chloroplast, the photosynthetic organelle in plants, much the same way an ancient bacterium evolved into a mitochondrion. The cyanobacterium lost the genes it needed to be freeliving and got busy turning water and sunlight and CO_2 into fuel and food for its master. All plants, animals, and fungi have mitochondria. But plants have chloroplasts, too. And that doesn't mean there were just these two endosymbiotic events. Many may have occurred. "Let me tell you a little secret," said Moselio Schaechter. "If you look, you will find bacteria that have bacteria-like bodies in them. The reasonable thing to assume is these capture events happen repeatedly, but the majority of these events don't lead to anything. They impart no advantage and so don't survive natural selection."

This kind of mucks up the traditional approach to the tree of life, where one stem branches into all the marvels of multicellular creatures. Professor Naeem was practically gleeful when he introduced us to the concept that the tree of life is not so much a tree as a bramble. He showed us a slide of the evolutionary biologist James Lake's proposal for a new tree of life. From a starting point springs two branches, bacteria and archaea, and then the two branches come together, essentially closing a "ring of life." From where they merge, another branch, the eukaryotes, grows, and from eukaryotes, plants, animals, fungi, and the protists, a grab bag of organisms defined by what they are not: not plants or animals or fungi. I noticed a tiny picture of Dr. Lake in the corner of a large illustration of this most untree-like tree of life. (A new tree of life published in the journal *Nature Microbiology* that looks like one of those spin paintings you can make at a carnival similarly shows eukaryotes emerging from a line where bacteria and archaea meet.)

Throughout his slides in class and in our textbook were pictures of the scientists who had solved riddles of nature, but they really all looked the same: either yearbook-type portraits of men or women in glasses or fuzzy field shots of biologists kneeling in the grass. Every scientific presentation I attended showed a slide that included the names and often little passport-size photos of the scientists and students who collaborated on a research project. I'm glad this happens. It's important to put a face to the experiment. I think we relegate those who speak the language of biology or chemistry or physics to a kind of rarified ghetto, where "they" have discovered liquid hydrocarbon seas on Titan, or "they" have discovered a protein used by bacteria to excise viral genes, but we can't really understand how "they" have done it because we don't really understand "them." Sometimes I think the use of the plural pronoun is a sign of our disassociation with the practice of science.

But it *is* sometimes difficult to connect to individual scientists. I went to an International Symbiosis Society (ISS) conference in Warsaw, Poland, in July 2013. ISS promotes the integrative discipline of symbiosis by connecting researchers. Every day I went to mind-numbing talks on subjects like the "Effects of aluminum and norflurazon on isolated algae from green hydra and related free-living species." During coffee break I hung around the cookie trays trying to start up a conversation. "What are you presenting?" I asked, and a young scientist launched into something about deciphering protein-coded molecular language, and I just smiled and nodded until it was obvious I didn't understand anything she said. In the inevitable awkward silence that followed, we inspected the cookies until finally, thankfully, we drifted apart.

Joining me in Warsaw just for the fun of it was a friend, Susan Murrmann, a miniskirt-wearing gynecologist who spends most of her week doing hysterectomies in Memphis and as a result is always up for a vacation. I was dutifully attending these presentations that were way over my head when after a day or two Susan demanded I find someone interesting to join us for dinner. During the lunch break, I plunked my stuffed cabbage down at a table where only a few people were seated and struck up a conversation with a boyish-looking fellow about food. Susan and I were staying at a swanky hotel with a decent restaurant, so I invited him and his wife to join us for dinner. We drank a lot of wine and ate beets in numerous manifestations and talked about current events in the USA. I didn't realize until years later, when I was looking at Professor Naeem's slides of the revised tree of life, that I'd shared supper with James Lake. It was a little like realizing you just shared a cab with Elvis.

The tree of life model has organisms diverging over time, but the endosymbiotic theory suggests there's more to evolution, that organisms can be brought together in a way that leads to entirely new life-forms. It's been witnessed in the lab. The biochemist Kwang Jeon at the University of Tennessee found that over the span of 200 generations, an amoeba (a bit of freshwater cellular goo) with a bacterial infection eventually co-opted the bacterial parasites to the extent that neither the amoeba nor the bacteria could survive without the other. They had together become something new.

"If we care to, we can find symbiosis everywhere," wrote Lynn Margulis in *Symbiotic Planet*. And she did. Margulis suggested that eukaryotic cells were a composite of bacteria with different abilities, contributing unique parts to the cell like a corporation. She saw the eukaryotic cell as a result of multiple mergers, some hostile, some mutually beneficial, the first being the acquisition of motion. In her view, spirochetes, the speedy spiral bacteria, joined with a host cell to create cilia (brushlike hairs that move), the tails on sperm, and other sensory protrusions. There hasn't been independent evidence to back up this hypothesis.

Her view, that symbiosis is the driver of evolution, not competition, has not been roundly accepted. Despite evidence supporting the idea that collaboration is also a mechanism of evolution, most people still think about evolution in terms of competition for resources and advantages. The evolutionary biologist Stephen Jay Gould called symbiosis the "quirky and incidental side" of evolution. Gould was one of the scientists who proposed punctuated equilibrium, which describes evolutionary change as long periods of little or no change "punctuated" by relatively rapid change. Margulis argued that's exactly when endosymbiosis happens.

At Columbia, my professors weren't so quick to pooh-pooh any of Margulis's ideas. "Lots of people rejected Lynn Margulis's endosymbiotic theory when it first came out," said Professor Naeem. "And they were later proven wrong." Scientists will argue about the extent of symbiosis in evolution, what the author and physician Frank Ryan has called "Darwin's blind spot," but about the origin of mitochondria, there is no argument. It remains the seminal eukaryotic experience, at the core of who we are, of every daffodil, every chanterelle, and every girl.

Indeed, our family history starts with a marriage of microbes.

I was beginning to feel like I was making progress in understanding how life-forms are connected. I hadn't even taken my first test, and I'd learned that the same six ingredients just keep being recycled in different forms over and over, connecting all life, and all oxygen-respiring organisms exist thanks to ancient cyanobacteria. I'd already learned that all bacteria are connected to each other, like one gigantic organism, and that plants and animals and fungi exist because of a connection made between two prokaryotes, when one engulfed the energy-making capacities of another and so allowed for life to become more complex.

But I didn't get to feel good about what I had learned for long. I hadn't finished answering the questions on the first study sheet when the second rolled in, encompassing two more lectures and chapters, and while it was not as many pages as the first, it contained way more complicated questions, like "Euglenids can be free-living photoautotrophs, while kinetoplastids, like the trypanosomes that cause sleeping sickness, leishmaniasis, and Chagas disease, are parasites, yet all are euglenozoids. What is the difference between a mutualist and a parasite, in this case?" There were dozens like this. And a few lectures later the third study sheet arrived in our mailboxes. This was the shortest one, only two pages, which was a huge relief, but then a few days after that it was updated, unapologetically, with another page of questions and a few more columns of terms.

Laundry didn't get done. I stopped paying bills on time. Dinner was haphazard at best. My husband Kevin would come home from work in the evening, and instead of seeing me behind the chopping board, a glass of wine in one hand and butcher knife in the other, he found a dark apartment, the only signs of life a strip of light under my office door. Inside I was surrounded by shifting strata of papers and open books with sticky notes slapped on every surface, lit by yellow puddles emanating from the three desk lamps I had rigged. I swung from book to lecture slides to my own notes to the internet, filling in the study sheets with my miniscule writing, my office drenched in the perfume of anxiety and unwashed hair. Kevin would peek in and I'd say hello without looking up, and then he would disappear, to fend for himself for the rest of the evening.

I took the study sheets with me when I went to the toilet. I read them in the subway, while waiting to see the dentist, while standing at the counter eating my lunch, in every café, in every line at the bank and the post office. I was obsessed—I had to be, because the only way I knew to memorize the material was by repeating it over and over again, until after a while I lost the meaning of the words and they became like a chant or incantation: Chromalveolata, Rhizaria, Archaeplastida, Excavata, Amoebozoa, Opisthokonta, pseudopods, filopods, axopods.

At night, I dreamt weird mash-ups of what I was learning: DNA reading frames on the loop of Henle (a tiny part of a tiny part of the kidney); mitosis of chemicals; segmented larva made of hydrogen bonds. In my dreams, I spoke biologese, but by morning I was illiterate again. The material penetrated every part of my consciousness, but I knew it was worth it because, so far, everything I learned about nature had a microbial connection. And I was beginning to think maybe the microbial way of life was the secret to understanding it all.

CHAPTER

5

Microbes Make Landfall

he first test of the semester covered everything from the origins of the universe to the basics of cell biology and the earliest organisms, and during the week before the test, our TAs offered test preparation sessions. I went to Yi-Ru's. She is a gentle woman, very open and friendly. Our study prep was held in a classroom and about six students showed up. Yi-Ru asked what we would like to review. A smiley junior named Austin said, "Why don't you ask *us* some questions?" The rest of us just plowed right over this suggestion and pummeled Yi-Ru with demands for explanations of quarks and plasmids. The problem with this approach was we all had different holes in our knowledge, and it was a waste of valuable study prep time to listen to an explanation of something you didn't need to prep for at all.

Later it occurred to me that Austin was both smiley and wily. Had we followed his lead, Yi-Ru would most likely have asked us questions that were at least similar to those on the test. One gal worked the pressure, begging for details. I didn't mind. Even though I wasn't competing with the other students for a position in grad school or angling for a plumy internship adding fluorescent tags to DNA in a lab, I did, in fact, want a good grade. In truth, the idea of failing appalled me. I mean, every time I passed my advisor, Matt, in the hall, he'd ask how I was doing in a friendly way and I would start worrying that he knew I was a chemistry dropout. The thing about biology tests is that unlike an essay question on, say, feminism in Hemmingway's novels, you can't wing it. So, yes, I hoped the gal who begged for clues as to what was on the test would succeed. Yi-Ru squirmed. Of course she wanted to tell us. Of course she couldn't.

At the end of the session, we clustered around a diagram depicting the six supergroups of eukaryotes and shared our techniques for memorization. This was pretty counterintuitive. You'd think all the formless goopy critters would be grouped together, but actually we are alarmingly close on the tree of life to amoebae. An Italian gal had a good grasp of Latin and Greek and she used it to remember that Rhodophyta were red algae (from the Greek for "rose" and "plant") and Chlorophyta were green algae (from the Greek for "green" and "plant"). Another girl preferred remembering the pictured examples in our book, though when someone pointed out there wasn't a whole lot of difference between the pictures of an euglenid and a ciliate (both of which look like pumpkin seeds), she blushed and said, "I'm doomed."

Memory used to be pretty important, but our technology has undermined the necessity for retaining knowledge. I don't even know my son's cell phone number, which—on nights when he was a teenager out partying and I was home worrying—I might dial three or four times, and still I never learned it.

Even when I was young, memorization was not my strong point, and it was one of the things that made me most nervous about going back to school. The Latin binominals of single-celled creatures? Please. I am prone to forgetting the names of people I meet within seconds of meeting them. I've even forgotten the names of my dearest friends at the moment of introducing them to someone else, to both their and my alarm. According to an article in *Psychology Today*, nearly 85 percent of middle-aged people forget names, and most of us worry it's a sign of early-onset Alzheimer's. However, the pamphlet *Staying Sharp: Successful Aging and the Brain* published by the Dana Foundation (a private organization that supports brain research) confirmed it is not unusual to forget names or where you parked your car. You only need to be worried if you forget you drove your car there in the first place.

Forbes magazine recommends the five best tricks to remember someone's name: Meet and repeat is number one. I remember meeting a politician in New York who repeated my name once or twice in the course of our brief exchange. It was pretty flattering at the time. But now I realize he was using the meet-and-repeat technique. Other recommendations are to spell the name out or associate the name with something personal. However, the article suggested that the main reason why we forget a name is that we don't really care. Which in a way is worse news than having an aging brain.

I did care about remembering all the terms in my bio book, but on the other hand, when I encountered something like "How homologous chromosomes find each other and become aligned is one of the great mysteries of meiosis," I was relieved. It was one more thing I didn't have to memorize.

Most students use word mnemonics to aid memory. That's where the first letter of words in a list are combined to form a phrase, and they worked for me: Please Please Mary A Taco for the stages of mitosis (prophase, prometaphase, metaphase, anaphase, telophase); All Girls Cut Time for the nucleic acids in DNA (adenine, guanine, cytosine, thymine); All Girls Cut Up for the nucleic acids in RNA (adenine, guanine, cytosine, uracil); and my opus, Come On, Stop! Dogs Can't Pee—Try Just Culling The Quart, representing the periods of natural history (Cambrian, Ordovician, Silurian, Devonian, Carboniferous, Permian, Triassic, Jurassic, Cretaceous, Tertiary, Quaternary).

There are also mnemonics you can choose from in case coming up with absurdisms isn't your forte. Learning the order of taxa (domain, kingdom, phylum, class, order, family, genus, species)? Let the experts recommend Dumb Kids Playing Catch On Freeways Get Squashed, or Dirty, Kinky People Can Often Find Great Sex. Taking the bar? Just remember MRS BAKER for common felonies (murder, rape, sodomy, burglary, arson, kidnapping, escape, robbery). Want to be a certified public accountant? To help remember the Code of Professional Conduct's principles (Integrity, Scope and Nature of services, Objectivity and independence, Responsibility, The public interest, Due care), try I S N O R T D.

I've noticed people really want you to use *their* mnemonics, kind of like how people really want you to use *their* dermatologist. I'd get a good one going: Get Some Guts, Maybe Cash, but our TA would recommend Go Sally Go Make Children for the five stages of the cell cycle (Gap phase 1, Synthesis, Gap phase 2, Mitosis, Cytokinesis), and then I wouldn't be able to remember either. Dueling mnemonics makes things worse.

Books on improving your memory are a market staple, with *Moonwalking with Einstein* by Joshua Foer being at the top of the list. The book starts with the story of the ancient Greek poet Simonides of Ceos. Moments after he stepped out of a banquet hall, the ceiling collapsed, killing everyone inside. Simonides reconstructed in his mind the banquet hall he had just left, and all the people in it, and so was able to point relatives to the location where they might find their loved ones' bodies. Simonides used an image mnemonic. This is the technique that competitive memory athletes use, and it's one that's been around for about 2,500 years. It's called the mind (or memory) palace, the method of loci (Latin for "places"), or the journey method.

Here's how it's done. You take a familiar place, like your gym, and choose a path, like lobby, weight room, spin room. Associate the information you are trying to remember with pictures, the gnarlier, the more memorable. For example, if you are trying to remember the functions of proteins, you could imagine transport proteins as guys pushing around carts of bloody towels in the lobby (the blood makes it more memorable), defense proteins as toughs pumping cannons in the weight room, and spin instructors as enzyme catalysts screaming, "Catalyze, people! Catalyze!" But if you are short on creative ideas of your own, SketchyMedical produces memory palaces for you. Med students that want to remember the details about the bacteria that cause cholera can watch a YouTube video of a drawing called Colonel Cholera's Base cAMP, which sketches a landscape and characters to remember details like the bacterium's shape and the illness's symptoms.

Short-term memories, like recalling where you parked, may activate neurons in the prefrontal cortex of the brain, right behind our foreheads. But long-term memories are processed in the hippocampus, a very ancient part of the brain.* When we use mnemonics or other memorization tools, we reinforce memory by passing the information through the hippocampus several times. Declarative memories, the memory of facts and events, are stored all around the brain, which is why a stroke can destroy some memories and not others. But the hippocampus and other parts of the temporal lobe are involved in figuring out whether a memory is worth storing and where to store it. Storage usually takes place in the area most involved with the particular type of computation to be stored, so facts about words would be stored within language areas. The actual memory is a strengthening of connections between neurons, making associations and faster processing possible.

But as you age, those connections between neurons shrink. The volume of the

---

*The hippocampus is part of the limbic system, what some people call "the lizard brain" because while that's about all a lizard has for brain function, the nickname implies its evolutionary antiquity.

brain diminishes. Bloodflow declines and you can't remember what you had for breakfast. When I was young, we called that senility, and if your great-aunt blew her nose with 20-dollar bills, well, that was considered a normal, if costly, part of aging. But the idea of inevitable nuttiness is no longer accepted. According to the Alzheimer's Association, most dementia symptoms are caused by Alzheimer's disease or vascular dementia, which occurs after a series of ministrokes. And it's pretty common. Over five million Americans suffer from Alzheimer's now, and by midcentury, someone will develop the disease in the USA every 33 seconds.

This awful news has stimulated a rash of prescriptive hoo-ha. Dr. Oz offers a five-step prevention plan for Alzheimer's that calls for "brain boosters": taking the supplement DHA, a fatty acid found in fish oil and mother's milk; crosstraining your brain by putting your watch on a different wrist or brushing your teeth with the opposite hand; cutting back on stress (a perennial on every health list); practicing memorization, starting with the grocery list or your kids' cell phone numbers; and stimulating bloodflow to the brain by doing pushups. *Real Simple* magazine says get more sleep, do aerobic exercise to improve bloodflow, eat blueberries and kale, have a cocktail every day, take fish oil (forget *Ginkgo biloba*—that was yesterday's supplement), meditate, experience novel situations that make you think, avoid antidepressants and antianxiety drugs, and check to see if you are gluten intolerant.

But these recommendations are just good guesses. It's natural to lose some memory as you age. Wandering around looking for glasses that are perched on your head or entering a room with intent and then standing there like a dummy trying to remember why is considered normal because it doesn't affect one's overall ability to live a productive life. But there is one consistent recommendation for keeping your memory vital, whether it comes from daytime TV or the National Institutes of Health, and that is *learn new things*.

If learning new things really helps keep the Alzheimer's bogeyman away, I think going back to school will ensure I have a sharp old-lady mind, because up until the day before the test, we were still learning. Indeed, our last class took us up to the primordial slime, the first gunk that washed ashore and founded the terrestrial ecosystem. Life on Earth is 3.8 billion years old, and the first three billion years of it was spent in the water, for a pretty obvious reason: two-thirds of the Earth's surface is covered with ocean. Early lineages of photosynthesizing cyanobacteria may have collaborated with bacteria that had the ability to acquire necessary nutrients, like phosphorus and nitrogen, and maybe other organisms, as well, to create tiny self-sufficient lifeboats. (It should be noted that some types of cyanobacteria were multitaskers and could utilize atmospheric nitrogen, as well.) That kind of team effort is still evident in oceans that are oxygen rich but nutritionally poor, in super-clear water much like the ancient seas might have been.

As their numbers grew, aquatic microbes conjoined into sticky microbial mats, green muck that grew on the floor of shallow oceans where light could penetrate, or floated on top of the seas and fresh water springs. For billions of years these mats were the only game in town, some growing like carpets, and others building layers like a cake. They were dense communities that were home to extremely complex relationships, so complex that tiny changes in the mat environment, like a decrease in pH or an increase of one species or another, could cause an adaptive shift of the whole. In *Garden of Microbial Delights*, Lynn Margulis and Dorion Sagan offered this analogy: Imagine one billion people living together in lower Manhattan. Now imagine the cooperation it would require for everyone to get along. That's what microbial mats were, and still are: really crowded, highly organized dynamic communities that are otherwise known as pond scum.

To live, the microbes had to share resources. One bacterium's chemical waste—oxygen, for example—became another's sustenance, because even waste contains energy. I once ate at a sushi restaurant with its own walls and décor and kitchen that functioned inside another Japanese restaurant, an accommodation, I guess, to the extortionate rents in Midtown Manhattan. Likewise, the living situation in microbial mats was so tight that microbes made homes inside each other, including the capture of a bacterium by an archaeal cell that led to our mitochondria and allowed for the evolution of beefy oxygen-respiring eukaryotic cells.

The first eukaryotic organisms living in these mats were probably single-celled organisms called protists, from the Greek *protistos*, meaning "first of all," and they gave rise to every line of eukaryotes. Protists aren't animals, but they can be animal-like (think amoebae). They aren't fungi, but they can be fungus-like (think slime molds), and they aren't plants, but some are plant-like (think algae). The kingdom of protists is really diverse—it's kind of a potluck of organisms living all kinds of ways—but generally, plant-like protists make their own food, like photosynthesizers; the animal-like protists hunt bacteria and archaea and each other; and the fungi-like protists are rotters. They eat the dead.

A few years ago, my mother became very sick with an animal-like protist called *Babesia microti*, a deer tick–borne parasite and increasingly common coinfectant with Lyme disease. She survived, but it wasn't easy. These tiny organisms invade red blood cells, where they rapidly divide, packing the cell until it explodes and releases a hoard to raid more red blood cells. Babesiosis causes a host of symptoms, including anemia, which my father, a fervent believer in food as medicine, treated with calf's liver and oysters. The disease has been around for a long time, primarily as a pathogen of cattle. In fact, babesia, which infects cattle by means of a tick bite, is the pathogen that kindled Theobald Smith's concept of insects as disease vectors in 1893.

Most protists are, like babesia, single-celled and contain everything needed for survival; all the organelles, the different proteins and enzymes, and so on, though its life cycle may require a host. Many are tiny, glassy cells full of churning internal parts that capture or absorb unsuspecting bacteria that swim by, like droplets of sugar syrup with intention and appetite, or green photosynthesizing beads of different shapes that bask in the sun, the microscopic version of grass.

Our biology class was a lecture class, so we didn't get to use microscopes to investigate a droplet of pond water. But I became rather addicted to the website pondlifepondlife.com, where I watched ciliates swimming around like clear, hairy bumper cars; or euglenids, tadpolelike things, zip across the microscope's round field of vision with whipping tails; or formless amoebae creeping along by extending their cell membrane and pressing their inner gunk forward in search of prey to engulf. Some protists are colonial, composed of 16 or 32 or thousands of cells, and are a focus of study for what they may tell us about the origins of multicellularity.

But not all protists are microscopic. Some are multicellular and gigantic, like *Macrocystis pyrifera*, the giant bladder kelp, which can grow to be 150 feet long at a rate of 2 feet a day. Cellular slime molds spend most of their lives as single-celled amoeba-like things until chemical cues tell them to aggregate into a great swarm. Plasmodial slime molds are basically enormous single cells composed of a swarm of fused individual cells, resulting in a single bag full of nuclei. They can be pretty alarming. In fact, folks from a Dallas suburb in 1973 were so unhinged by a large specimen of *Fuligo septica*, which looks like living scrambled eggs, they feared it was an alien. Another slime mold, *Physarum polycephalum* (many-headed slime), has shown intelligent characteristics. It will find the shortest way through a maze in order to get at the food in the center. When researchers placed bits of food on a map of major Japanese cities, the slime mold formed connected branches that linked the food sources. The network had a remarkable resemblance to the country's rail system. "Working with [slime molds] constantly challenges our preconceived notions of the minimum biological hardware that is required for sophisticated behavior," said Simon Garnier, an assistant professor of biology at New Jersey Institute of Technology who ran a series of tests of slime mold decision-making. But mostly people know about the single-celled pathogenic amoebae, like the notorious *Naegleria fowleri*, the brain-eating amoeba ("They're a bit on the nasty side," said Tim Grabham, director of the 2014 slime mold movie, *The Creeping Garden*, in dependably English understatement), which infects people when contaminated water is run through the sinuses—one of many reasons why you should sterilize your neti pot regularly.

Studying the different phyla of protists was frustrating because they are such a random group and there are so many of them—biologists have identified over 200,000 species and there are doubtless many more—that it's super hard to remember them in a hierarchical way. It's like learning random numbers. The only thing that worked for me was the pictorial mnemonic. I made one for Rhizaria, a group of protists that use a footlike appendage to drag themselves around, based on the rooms in my first apartment, a dump above Mamoun's Falafel in Greenwich Village. But since there are six main groups of protists, after a while I got pretty desperate for blueprints and started mining my childhood home in upstate New York (for the Excavata) and my grandparents' condo in West Palm Beach (for the Chromalveolata).

I had a hard time keeping track of all the memory palaces I had created. It was like trying to cook 10 dishes at once, and the day before the test, I was surging with cortisol, a stress hormone. A study found that participants with increased cortisol performed poorly on memory of socially relevant information, like people's names. Anecdotally, I can confirm that increased cortisol negatively affects one's ability to remember protist names, too.

It's a feedback loop. Forget the name, pump cortisol, pump cortisol, forget the name. Cortisol sends glucose into your bloodstream when you encounter a stressful situation. It's useful when, say, you run into a yeti in the woods. The glucose gives you the added energy you need to either run away or put up your dukes. But it's not particularly useful when your job is to be still and recall something.

The effect of stress on performance, whether it's taking a test or giving a talk at your local library, is one of the most researched phenomena in psychology. A kind of performance anxiety, its reach is wide, from stage fright to erectile dysfunction. The Yerkes–Dodson law (developed by two scientists in 1908) shows when your stress is high, your performance tanks, though you need some stress to perform at your best. Freezing, blanking out, and not being able to breathe are all stress reactions. "With too much stress . . . you can't think through a question. With too little stress there's no juice. You blow it off," said Ben Bernstein, a clinical psychologist and author of *Test Success! How to Be Calm, Confident & Focused on Any Test*.

Gut microbes from the bacterial genera *Escherichia* and *Bacillus* and the fungal genus *Saccharomyces* produce norepinephrine, a stress hormone similar to cortisol. The hormone travels to your brain by means of the vagus nerve, which runs through your guts. Hormones can also affect changes in your gut microbiota, causing some species of bacteria to increase in numbers and virulence, and that can make you sick. Maybe that's why so many students become ill after exams.

By dinner the night before the test, I had a sore throat and felt mentally full. Not satiated, but overstuffed, like after Thanksgiving dinner. Every room of my memory palace was bursting with bizarre images. I'd crammed the protists into every crack and corner. My cortisol was raging. I was brusque with my husband when he tried to give me some test-taking advice. "Do the easy questions first," he said, "then the questions that garner the highest points." And I asked him, "When was the last time you took a damn test?" I was too wired to sleep, so I did what any sensible middle-aged person in my predicament would have done. I took a Xanax.

The next morning, I collected my study sheets, filled two mechanical pencils with lead, and got on the 1 train uptown. I gave myself a full hour to get to the test, just in case of a subway delay. I didn't mind the subway ride uptown. It takes about 30 minutes, and I usually spent the time reviewing my study sheets, mumbling the definitions under my breath, "endocytosis, phagocytosis, exocytosis, transcytosis," holding on to the subway straps with three fingers to minimize my exposure to whatever bacteria were lingering around. Fear of subway germs is old news in New York. In the fall of 1990, the *New York Post* featured a homeless junkie in an article, "Tuberculosis Timebomb," claiming he was spreading TB along the IRT line. More recently, in 2015, researchers at Weill Cornell Medicine released a study that mapped DNA found in the subway system. Twelve percent of the samples revealed disease-associated bacteria, including bubonic plague, all of which were dead, and the researchers later retracted some of those identifications. Nonetheless, this set off a rash of headlines among online news sites, like "NY Subway Has Bubonic Plague" (from Newser, whose tagline is "Read less, know more"), accompanied by pictures of crowded subway cars and rats on the tracks.

The intercoms on New York subways are highly variable, with communications from the conductor ranging from crisp and annoyed ("Do not hold the door, people. Step in or step out.") to mumblers whose instructions are completely garbled with static. It was just such a conductor that was driving the 1 train uptown as I headed to my test on 116th Street. As we approached 96th Street, an express stop, the intercom fritzed into life and the conductor said something in a scratchy, utterly bored voice. We passengers looked at each other in mild confusion, shrugged, and went back to our reading. And then the train picked up speed.

I know how fast a train goes between local stops, and this was too fast. We passed the 105th Street stop; we whipped by the 110th Street stop. As we barreled through Columbia's 116th Street stop, I started to get nervous. But it was okay, I told myself. I have plenty of time. I'll just get off at the next express stop and catch a downtown local. I tried to study some more, but once we emerged from underground and were racing along an elevated track in the sunshine, I couldn't act cool any longer. I stood up and began pacing the length of the car in search of a subway map free of scratchiti. When we screeched to a stop at 137th, I bounded out of the doors. I'll just catch a cab back downtown, I thought. No big deal. I'll make the test.

Manhattan is filled with yellow taxis cruising for fares in midtown, but at 137th? Fuggedaboutit. Green gypsy cabs rule the streets uptown, and while officially they can be flagged down, they're mainly livery cabs, and I've never had one stop for me. I walked backward south on Broadway, facing the oncoming traffic, practically *in* the oncoming traffic, but car after car passed. Ahead was a bus stop, and I could see a Broadway bus lumbering toward it. Great alternative! I'll take the bus downtown.

We made two stops, and then the driver opened the door, stood up, put on his sweater, and got out. I could see him chatting on the sidewalk with another driver. If you are in a hurry, a driver changeover is one of those maddening things. Because the new driver needs to get settled and stow his lunch bag and adjust his mirrors, and he's taking his time and you are bouncing in your seat thinking, Go! Let's go! But finally, the bus lurched into motion and I breathed a sigh of relief. About 10 blocks later, the bus stopped again, and in one of those cosmic irony-type moments, the driver opened the doors and began the agonizingly slow procedure of lowering the wheelchair lift.

I did make the test. I had to run up five flights of stairs, and I burst into the classroom with a pounding chest just as the exams were being passed out. I noticed my usual neighbor had moved to the front row at the end so that she would be the first student to receive the quiz.

It was 15 pages long, with 35 questions ranging from a quarter to three points. I trudged through it, from first page to last, trying to ignore the other students as they got up, slung their backpacks over one shoulder, handed in their tests, and left. There were all kinds of questions with tricky wording that I read wrong the first time, like "Which statement, if it were true, would mean

Earth would never have formed or it would be so different that it could not sustain life as we know it and we would not be here to take this quiz?" Others were so bogged down with jargon it was like wading through mud to read them: "Epinephrine and glucagon are both hormones that activate GPCRs whose G proteins activate adenyl cyclase. In conjunction with cAMP production . . . " Well, the swamp went on and on, a virtual Okefenokee. There were agonizing diagrams of molecules that I suppose the students who studied chemistry sometime in the last 5 years recognized, but not me. At points in the hour, I felt my mind wandering to my memory palaces, and with each turned page, I hoped I would encounter more questions I had studied than those I had not. It was devastating. When I finally turned in the test, I was the last person to finish.

I really didn't have much time to sulk over the test, because there were new chapters on multicellular organisms to read. Multicellularity is not to be confused with a colony of cells. Each bacterial cell in a colony is a separate organism with its own DNA. In a multicellular organism, all the cells have identical DNA, but the genes in each cell are turned on or off depending on what role that cell is playing in the body. It's the gene *expression* that varies between a cell in my eye, my hand, my heart.

The microbial mats that eventually nurtured multicellular life were complete ecosystems, really, the primordial ecosystem. Even sex evolved in mats, making it one of the most ancient practices on Earth. Bacteria, of course, predate sex. They exchange genetic material, but asexually and randomly. Matt Ridley, in his book on sex and evolution, *The Red Queen*, put it this way: Asexual breeding is like having lots of lottery tickets mostly with the same number. The chances of getting a winning ticket—in the game of evolution that would be a mutation that increases survivability—are pretty small. In the case of bacteria, those odds are improved by the speed of their multiplication, and the more bacteria there are in a given space, the more likely they will encounter one another and share useful genes.

What we do, in contrast (as do lots of other eukaryotic organisms), is produce offspring with unique combinations of genes that may afford some advantage in a changing environment, one generation at a time. To continue with Ridley's analogy, sexual breeding is like having lots of lottery tickets with all different numbers. The chances are higher of hitting the jackpot. In life's lottery, that would be some trait that helps an offspring survive. If there is enough genetic diversity in a population, natural selection will act on the fittest individuals and they will define the species going forward. Sex plus natural selection rids the genome of unfortunate mutations, whereas in asexual populations, deleterious mutations tend to accumulate. That's why they need lateral gene transfer and virus-vectored gene transfer to mix up their genes. "Sex," wrote Ridley, "is a sort of free trade in good genetic inventions."

Many scientists think eukaryotes evolved to have sex as a response to bacterial pathogens. Sex gives eukaryotes the genetic opportunity to out-evolve a predatory bacterium. The idea, proposed by the evolutionary biologist Leigh Van Valen, is called the Red Queen Hypothesis, and the name comes from a scene in *Alice in Wonderland*, where Alice is running a race with the Red Queen and her entourage but they aren't getting anywhere.

"Well, in our country," said Alice, still panting a little, "you generally get to somewhere else—if you ran very fast for a long time as we've been doing."

"A slow sort of country!" said the Queen. "Now, here, you see, it takes all the running you can do to keep in the same place."

The hypothesis proposes that just as the Red Queen has to run to stay in place, genetic recombination—sex—helps the organism stay an evolutionary step ahead of the parasites, predators, and competitors that are evolving ever new ways to wipe it out. This suggests human beings have sex to maintain a kind of détente with our traditional antagonists, bacteria. They adapt to get around our defenses, and we in turn adapt to combat the new lethality. It's a biological arms race, a kind of cold war conducted sexually: You don't win, but you don't lose either.* You could say bacteria are the reason why we have sex in the first place.

---

*It's arguable that access to health care is a form of artificial, or man-made, selection, because those who have it will go on to reproduce, whereas those who don't may not. That's a heinous thought.

Professor Naeem described microbial mats in the oceans as pastures that supported the evolution of grazers and burrowers, and I imagined green fields full of munching cowlike crustaceans. But actually, it was probably more like the slimy, stinky crusty stuff that coats a bay floor at low tide. And the grazers might have been the curious *Ediacaran biota*, which look like fleshy ferns and flat sink stoppers, most only a few centimeters around, but some as large as a bath mat. They also look bilateral, which represents an evolutionary advance. Virtually all animals have either radial symmetry, where an organism can be divided neatly in half along any axis, like a starfish, or bilateral symmetry, where an organism can be divided neatly in half along only one axis. We are bilateral. We have a right half and a left half and a front and back. Being bilateral was the innovation that allowed animals like us to move forward through the environment in a consistent direction, which in turn led to a grouping of nerve cells into a brain, with the brain's sensory structures, like eyes, in front.

The microscopic burrowers were worms, and they represent an evolutionary advance, as well. In order to burrow, they evolved a coelom (pronounced see-lum), a body cavity, and this is a key feature that separates animals from relatives like sponges. Here's how our teacher explained it. What happens when you bend a hot dog? It breaks, right? That's because the inside is stuck to the outside. A body cavity, however, allows guts to move around so an organism can bend. If it can bend, it can burrow. The burrowers ploughed the mats and the compacted oxygen-deprived sediments below, which ventilated the sea floor and allowed it to become habitable to more oxygen-respiring organisms. This is called the agronomic revolution. Agronomy (from the Greek for "field" and "law") is the scientific study of soil management. Which means the first farmers were worms that ploughed microbial mats in the Cambrian seas.

This was the time of the Cambrian explosion, an era of diversifying lifeforms, though the Cambrian fossil explosion might be more accurate. There may well have been other "explosions" of life, but most organisms don't become fossils. The circumstance by which a fossil is preserved is rare, and our understanding of evolution is limited by preservation bias. But what we do know, we

know because of the Burgess Shale, layers of ancient sea floor mud that accumulated 505 million years ago.

What is unique about the Burgess Shale Formation, which surfaces in the
Yoho and Kootenay national parks in the Canadian Rockies, is the collection
represents all major body plans we still see in modern animals today: starfish,
worms, octopuses, scorpions, sponges, and chordates, animals with a hollow
dorsal nerve cord and a post-anal tail, which includes us. The Burgess Shale
records the weird youth of animals, including bizarre creatures like *Hallucige-
nia*, a walking worm covered with thorns, and *Odontogriphus*, which look like
a Dr. Scholl's insole with a mouth. There's not as much known about Cambrian
plants. They didn't come up in class; we were all agog over the weird illustrations of the fierce 3-foot-long *Anomalocaris* with its round tooth-rimmed
mouth, which, ironically, may be the ancestor of the shrimp. Cambrian flora is
thought to be mostly small, unicellular or branching, soft, mucky green things.
Plants would have their day once they moved to dry land.

As the ocean population diversified, it created a more complex ecology,
which supported further diversification based on either finding new ways to eat
things, by evolving a new mouthpart, for example, or finding new ways to avoid
being eaten, hence the proliferation of fossils with all manner of spines and
horns and convoluted shells. All parts of the oceans were used—the sea floor,
the sediments below, and the water column above. Some marine animals even
ventured onto the beaches. There are fossil prints of lobster-size, woodlice-type
things scuttling over the sand like crabs exploring the low tide. But these
weren't terrestrial pioneers, just visitors.

In order for animals to make it on land, there had to be soil first. The con-
tinents and islands of Earth were bare rock and sand, carved by wind and rain.
There was no soft, fertile place for life on land to get started, so life came ashore
with its own fertile ecology. Microbial mats made landfall during the Ordovician, about 475 million years ago, teeming with bacteria, archaea, protists,
probably fungi, and the scummy green forefathers of plants. No one particular
creature made landfall on Earth. A whole ecosystem did.

Over time, those microbes that acquired the genes necessary to survive
longer out of water prevailed. Evolution was going like gangbusters in the
oceans, but on land, it had to start almost from scratch. Moving to land had its

allures. In contrast to the oceans, there was lots of sunlight for photosynthesizers, lots of free real estate, and no competition. But the challenge was finding ways not to dry out. And, indeed, the legacy of our aquatic past is all organisms on land evolved to hold on to as much water as they can. We still start life in a protective sac of water.

I've seen something like those ancient mats growing on sand dunes on Long Island. They're called cryptogamic crusts, biological communities composed of living and dead mosses, lichens, algae, fungi, protists, and bacteria that cover arid ecosystems. They're a crumble of intermingling organisms that look like streusel resting on the land. When I stepped on it, it crunched as if it was dead, but it also felt unholy, like grinding your foot into an anthill. And, indeed, a close look would reveal in miniature the chemical workings of the entire living planet: the production of oxygen, the fixing of nitrogen, the cycling of phosphorus and carbon. This is what the primeval soil, the protosoil, looked like, what the mycologist Nicholas Money calls a "crème brûlée," a crusty-on-the-top and soggy-below mix of organic and inorganic elements like sand and clay particles bound together in a voracious food web that spread over the earth like skin.

When microbial mats made landfall, they brought with them a complete ecology that included the seminal food web the planet would need to transition from rock to garden, that allowed for the flippery fish from my childhood books on evolution to crawl out of the ocean to find food, that made Earth habitable for us. As paleontologist Richard Fortey wrote in his book *Life: A Natural History of the First Four Billion Years of Life on Earth*, "Mats maketh man."

CHAPTER

6

The Soil Microbiome

I was pretty apprehensive during the week or so we had to wait for our tests to be returned, so I appreciated the way the TAs finally handed them to us—after class, expressionless, and folded. And I was even more relieved to learn I'd received 17 out of 25 points. I had expected worse. I looked over the questions right away, trudging down the five flights of stairs while students winged past me like bats in a forest. Rats! I had completely over-looked a full-page multipart question concerning how the polymer of glycogen is broken down to monomers, worth two points. I couldn't remember if I thought I'd go back to it because it was complicated or missed it in my anxiety, but once home, I answered the question just to see how I would have done. I got eight out of 10 parts correct. Where I am at in my education, it was enough to know I'd learned the material. But then I noticed the first question. What is an example of an emergent property? And I wrote "human consciousness," which was marked wrong. An emergent property is a result of the way compo-nents react that often can't be deduced just by looking at the parts themselves. Leonardo's heart plus lungs plus stomach equal the Last Supper? I don't think so. I decided to grub.

Grade grubbing usually occurs at the end of a semester, when students are looking to bump a B- to a B. It's common, and teachers dread it. On the Chronicle of Higher Education website, there's an on-and-off-again thread called the "Grade-Grubbing Hall of Fame," where professors post some of the pleas they've received, from "My grandmother was in the hospital the day of the final" (one teacher heard this from two students in the same class) to this (slightly edited) exchange between a biology teacher and student in 2011.

Hello, Professor: I am very distressed to see I got a C. I am from India as you know, and the stresses on girls in my country are tremendous. If I do not get an A, my future will be ruined. Please, sir, can you change my grade to an A?

The response was the standard "you earned what you earned" line. But the student was persistent.

Hello, Professor: If I do not receive a grade change, my family will send me back to India and marry me off. And women have very little opportunities in India, especially once they are married. I am pleading with you. Can you please grant me a second chance to do well and succeed in America?

The professor's response: "Please consider the issue closed." She did, but with one parting swipe.

Fine, Professor: But please understand while you are sitting high and mightily in your office that you have condemned me to a life of misery and drudgery washing out my husband's underwear in filthy water. I think you are a horrible teacher, anyway.

I wrote an email to Yi-Ru about the emergent property question, and she gave me another quarter point. I also mentioned that I'd finished the missing question at home, but I didn't ask for the points outright. It was more of a passive grub, as it crossed my mind she would assume I was too mature to cheat and so the point was legit. But she, in turn, responded passively, too. She pretended I hadn't mentioned it at all.

During the next few weeks, the class veered into animal organ systems: osmotic regulation (peeing, in short, or in the case of birds and reptiles, squirting white uric acid), the respiratory and circulatory systems, the muscles and skeletons. While it was satisfying to learn more established science than, say, the origin of the nucleus, I was impatient with this material. It had nothing to do with microbes, and learning about the workings of the animal body seemed mostly about learning the names of parts of the body, and that was stupefying: the enkephalins and endorphins, the nerve rings and nets, the chondroblasts and chondrocytes, and the osteoblasts and osteocytes . . . it was like picking up a spilt box of matches one match at a time. I wanted to learn about the microbial communities on Earth. I wanted to learn about soil.

But maybe the reason why it wasn't on our itinerary was because soil science is complicated. It's a microcosm of its own. Five hundred years ago, Leonardo da Vinci said we know more about the movements of the celestial bodies than about the soil underfoot. And it's still true. The identities and roles of 99 percent of soil microorganisms remain a mystery, and even the relevance of their diversity is unknown. How many types of microbes does it take, what kinds of jobs do they have to do, for soil to become soil and stop being dirt? Scientists involved in efforts like the Earth Microbiome Project and TerraGenome are trying to figure out who all the microbes are by sequencing genes in particular soil ecosystems, like the Mediterranean grasslands. But the problem with soil metagenomics is you can identify genes but that's not the same as understanding what those genes do. You could sequence all the billions of genes in a teaspoon of soil, and you could say, okay, here's a methane-generating microbe, but that doesn't mean it is actually generating methane. It may be capable but only actually generates methane if there is a change in, say, oxygen levels. And maybe that change is predicated on the presence of another microbe altogether.

In the mucky middle of my first term, I decided to find out. Anyway, I'd just had it with repeating the different types of tissues over and over. It was time to take a little vacation. To Kansas.

Every year the Land Institute, which is based in smack-center-of-the-state Salina, Kansas, throws a weekend event with lectures, musical performances, BBQs, a bonfire, and tours of the institute's crop-breeding plots. The Land Institute is a science-based research organization that focuses on the development of perennial—that is, constantly growing—strains of edible grains (like cereals, oilseeds, and legumes) in order to capitalize on the natural resources of unmolested soil. The annual Prairie Festival is a combo soil-science seminar, with updates on the breeding programs, and celebration. It's what cofounder Wes Jackson called the "intellectual's hootenanny." I drove into Salina at about 90 miles an hour. I almost missed the turn off because the road from the Wichita airport was so straight I had no idea how fast I was driving. Indeed, driving through miles of wheat fields was as lulling as Muzak. The biggest objects on the horizon were lumbering yellow tractors, like elephants on the savannah. The institute is just outside town, composed of a cluster of modest beige buildings that remind me of the Mennonite farms I've seen in Colorado, tidy and no-frill. There is a research building and lab and a long greenhouse or two, and around the campus are fields where the institute scientists are growing test crops.

I registered in the kitchen of a split-level ranch house and made my way to the reception area where I joined a crowd nibbling teeny Swedish meatballs (Swedes were a major immigrant group in Kansas in the latter 19th century). My fellow revelers included a mix of advocates—the Beltway types, who if they don't know everybody, seem like they do, and the homegrown types, people that fought for civil rights when they were young and are back at it, fighting for Earth's rights now. I observed scientists navigating the art of small talk and folks like a senior couple from Athens, Georgia, who spent a lifetime bringing a beat-up piece of land back to productive health so they could pass it on to their kids. There were young men in beards and buns and girls in bare legs and sandals and academics, bookish and rumpled, and talking to everybody was Wes Jackson. Donning blue jeans and homespun idioms, he's an affable and frank intellectual, a kind of Aristotle of the prairie.

Lectures started up promptly the next morning. In contrast to the silent snoozers and surreptitious cell phone users in my biology class, this audience was highly engaged, quick to applaud, enthusiastic in their oohs and aahs. The institute's goal is to transform farming from annual monoculture to perennial polyculture, an agriculture that functions more like the ecosystem that agriculture replaced. In order to understand why that goal is important, you've got to understand the ecology of soil. And that's what the speakers, from academia, the Union of Concerned Scientists, and the United Nations, set out to do in an open-sided pole barn where wasps floated in the still air of the eaves. There's not as much soil on Earth as you'd think. If Earth is an apple and you peel 75 percent of the skin off, that would represent the water. Now peel another 15 percent of skin off, and that represents the barren places, like deserts and mountains. The 10 percent remaining represents all the soil that sustains all the plants that sustain us all.

Soil varies. Temperate grasslands have rich soil that stores nutrients, but tropical rainforests, in contrast, have nutrient-poor soil. Tropical forests don't experience seasons; they're evergreen. A heavy canopy ensures a dark and moist forest floor where microbes decompose the constant leaf litter. Nutrients don't get sequestered in soil. They get used right away. But wherever soil is, however much there is, soil is composed of living and dying microbes. It's a more mature version of the microbial mat—nature's threshold ecology. Without microbes, soil isn't soil; it's sand or clay or silt or rock. With microbes, sand and clay is loam.

And there are a lot of microbes in soil. One handful of soil contains 50 to 100 times more microorganisms than people on this planet. That's 10 billion to 50 billion bacteria, hundreds of millions of viruses, 100 million different fungal cells, several thousand protists, hundreds of insects, and much more. Another calculation showed the total mass of organisms under an acre of wild land was equal to at least 10 draft horses. The numbers, wrote the authors of *Trees, Truffles, and Beasts*, "are as difficult to comprehend as the national debt." And all of them are caught up in a complex food web where everybody needs helpers in order to make food or to eat someone else.

All actors in the theater of soil are important. Even though fungi evolved over a billion years after bacteria, they make up for their late entrance by the key roles they play. Fungi may well have been the first types of organisms to emerge from the microbial goop that accumulated on ancient seashores, yet they don't get very much cred. Despite a yearlong biology course, there was no reading on the subject. We spent three slides in one lecture on fungi from an evolutionary point of view and three slides in another on fungi from an ecological point of view, and that was it. Being a student of biology, I was a little disappointed. Being a mushroom enthusiast, I was very disappointed.

Most of what I learned about fungi came to me via the New York Mycological Society, which was founded over 50 years ago by the composer John Cage. He set up the club's only rule, that there are no rules. The club meets on weekends to hunt mushrooms in the city parks. You can spot them on an October afternoon with their baskets and walking sticks, cheerily tromping around Woodlawn Cemetery in the Bronx, checking under the old oaks for wild maitake mushrooms. The club's lead scientist, Gary "the Woody Allen of Mushrooms" Lincoff, is famous in mushroom circles for his snow day hunts where red-knuckled collectors vie to find pieces of wood covered with rusty gunk that looks like dried vomit. But a walk among the trees of Central Park with Gary has taught many New Yorkers basic fungal literacy, and that's important if you want to understand the nature of soil.

Of the six kingdoms of life on Earth (eubacteria, archaebacteria, protists, fungi, plants, and animals, to use an old-fashioned kind of taxonomy), fungi are among the three most complex, along with animals and plants. They are eukaryotes that can be unicellular, like yeasts, or multicellular, with different cell types. They are mostly small—a fungal cell is only three to five times bigger than bacteria—and can grow in continuous branching tubes or long chains of cells joined end-to-end called hyphae that are at least 10 times thinner than a human hair.

Their reproduction may be asexual or sexual (mushrooms, which are like a flower or fruit, are the organs of sexual reproduction of about 20,000 species). Fungi are heterotrophs, meaning they eat other organisms like we do. In fact, they are more closely related to us than they are to plants (which is why it is so hard to beat a fungal infection—the medicine that kills the fungus can hurt us, too). They specialize in extracting and absorbing nutrients from their environments, but because they are too small to have stomachs, fungi do their digestion outside their bodies. They release the equivalent of gastric juices from the tip of their hyphae; degrade the bonds between complex molecules like fats, carbs, and proteins; and consume the smaller molecules those fats and carbs are made of. The ecologist Suzanne Simard once helped me understand it this way: If you were a fungus that wanted to eat a steak, you'd lie on the beef and your stomach acid would seep out and break down the meat, and then you'd absorb the nutrients back into your stomach. As long as you had steak to eat, you could keep growing, which is why fungi are among the largest known living organisms (an endless steak could feed an endless fungus).* They also don't have to die eventually like us. If circumstances allowed, it's possible they could live forever.

All fungi secrete enzymes to break down their food, though they don't all use the same ones. Each fungus produces the enzymes it needs to break down the foods that are part of its evolutionary niche. It's not dissimilar to animals, where some species have the enzymes to digest caribou, others, bamboo. As the fungus grows, the hyphae branch and rebranch, spreading in every direction it has food to grow into. That's why a mushroom ring in your lawn or your dog's ringworm has a circular pattern. The fungus is growing out from its point of origin in every direction it can and at about the same overall rate. In the case of fungi that produce mushrooms, the mushrooms fruit along the circumference.

When you kick open a rotting log, that cottony stuff inside is actually

---

*The largest and one of the oldest organisms discovered is the "Humongous Fungus," a wood-decaying *Armillaria gallica* that produces the edible honey mushroom. This giant mass of mycelium lives in the Malheur National Forest in eastern Oregon. It is 2,200 acres large and at least 2,400 years old.

millions of fungal hyphae that have massed into a mycelium. There are thousands of pounds of this stuff in every acre of soil doing all sorts of jobs. They
are decomposers, called saprobes, which eat dead and dying things. Without
them we would be buried under miles of dead plant debris. With them, the
carbon and other nutrients in dead plant matter are recycled back into the soil.
And they are mutualists with plants, trading nutrients they find in the soil for
sugars and fatty acids produced by the plants. There are a few types of mutualists. The mycorrhizae are fungi that live on and in the roots of most plants, and
the endophytes live between or within the cells of plants—all plants. And
finally, some fungi are parasites, which kill plants and other organisms, especially weak ones. Sounds tidy, right? It's not. A mutualist can become a parasite
if the circumstances dictate or become a decomposer if the host dies. Then
things get really confusing.

Not only are fungal lifestyles capable of transitioning from one type to
another, but there are a lot of fungal species, as many as 3.8 million including
lichens (that's nine times more than plants), but only a tiny percent have been
named so far, and among that number, how the different species are related to
each other is an ongoing hash. You can't use the biological species concept (who
has sex with whom), because they won't reproduce in a petri dish and it's not
something you can watch happening underground. You can't use the morphological species concept, because, like many microscopic organisms, they tend to
look a lot alike. As a result, species are verified by genetics, and that has led to
a reshuffling of mushrooms into new genera, so much so that mycologists complain it's impossible to know what is related to what anymore. "It's like having
early-onset Alzheimer's," said Gary Lincoff. "I know fewer mushroom names
every year."

That said, according to my biology book, there are seven fungal phyla,
based on how they produce sexual spores. Keep in mind, a phylum is a very
wide category. To remind you of how wide, ours is Chordata, and it includes sea
squirts and the queen of England. Just as in the different phyla of bacteria, there
is a lot of diversity within each fungal group and almost all of them contain
decomposers, mutualists, and parasites.

It was members of the phylum Glomeromycota that likely made the evolution of terrestrial plants possible. Their hyphae penetrate the root cells of plants

by chemically suppressing the plant's immune system, which allows the fungus to deliver an IV of mineral nutrients to the plant. These are endomycorrhizae (from the Greek for "inside," "fungus," and "root"), and they are present in 80 percent of all plants. These fungi don't produce mushrooms. Ectomycorrhizae (from the Greek for "outside," "fungus," and "root") do produce mushrooms. Their hyphae concentrate in the topical root cells in order to deliver their nutrient payload. Species within the phyla Basidiomycota, Zygomycota, and Ascomycota form this kind of mutualistic relationship. Ectomycorrhizal fungi partner with fewer species of plants, about 3 percent of mainly forest trees, but they are ubiquitous because their host plants exist in huge numbers.

Endo- and ectomycorrhizae increase a tree's access to nutrients 1,000 times beyond the reach of its own roots. Indeed, if you brush away the soil from the roots of a conifer and squeeze the tips, they'll smell like mushrooms. They are the "pipelines of the environment," said Kristine Nichols, the Rodale Institute's chief scientist. They nurture tree seedlings by colonizing the young roots and connecting the seedling to the nutrient infrastructure of mature trees. Mycorrhizae are the bosoms of the forest, connecting a wet-nursing tree to a hungry seedling.

Fungi in microbial mats may have initiated soil building by mining the primeval rock for nutrients and decomposing dead organisms in the mats to capture their carbon and then diversified into mycorrhizal roles as plants evolved. The earliest terrestrial fossil found so far is a fungus called *Tortotubus*, which looks like a microscopic split end and colonized Earth 440 million years ago. Likewise, lichenlike organisms may have been early pioneers. There are fossil lichens dating from 400 million years ago.

Lichen is a partnership where photosynthetic algae or cyanobacteria are housed within a fungal web. Imagine the cyanobacteria as beads and the fungus as a kind of delicate macramé of one-cell-thick threads binding them into position. In this relationship, the cyanobacteria get a safe haven from predators—a really safe haven, as lichens can withstand everything from drought to outer space—and they get water, too, since lichens can hold three to 35 times their weight in water and take a really long time to dry out. In turn, the fungus, which doesn't make its own food, is provided sugar by its in-house chef.

Modern lichens have bacterial symbionts, and ancient lichens probably did, too. These bacteria may have served a number of roles, like providing vital nitrogen to both the fungus and its photosynthesizing partner and freeing up carbon and other nutrients by decaying the dead. And the symbionts don't stop there. It turns out that leafy lichens are covered in yeast, which seems to play a role in producing specific degrading acids. Lichens grow up and out, but underneath they decompose, building soil at the same time as the fungal parts secrete acids over rock, degrading it like hot water over a sugar cube, prospecting for nutrients. Their tiny rootlike probes are tough enough to pull rock fragments apart but sticky enough to glue the loosened particles together. Fungi were the first engineers of soil.

Primitive plants were evolving at about the same time. Aquatic algae are the ancestors of land plants. They had the genes to interact with beneficial microbes in the water, and it is quite likely their ability to utilize bacteria and fungi to acquire nutrients in water allowed them to colonize land. In fact, you can see those partnerships in liverworts, perhaps the earliest land plants.* Modern liverworts use mycorrhizal fungi to help with their nutritional needs, and ancient liverwort fossils reveal fungi lived between their cells and in their roots, too. And modern liverworts are speckled with nitrogen-fixing bacteria. It's not a stretch to think ancient liverworts were as well.

Liverworts lack the features of more modern plants, like stomata, a kind of venting apparatus, and connective tissues that make stems, but they have a waxy coating that protects them from drying out. That was key in the early days of landfall because the two main challenges plants faced were how to retain water and how to deal with gravity. When you look at liverworts, which grow at the water's edge, they represent a stage in plant evolution when their waxy skins had solved the desiccation problem, but the gravity problem remained. Indeed, liverworts aren't upright. They have flat leaves that spread out over wet soil like green coins scattered across the mud, and they are anchored into the earth with primitive roots that look like sticky whiskers.

During the Silurian period (450 million to 400 million years ago) plants acquired roots, stems, and leaves, as well as vascular tissue that allowed nutrients and water to travel up a stem, which meant they could increase their access

---

*They are so named because the leaves are shaped like a liver and were thought to cure liver disease, a notion based on the doctrine of signatures (developed by a Swiss alchemist in the 1500s), which says that God made plants look certain ways so we would know how to use them.

to sunlight by growing up and leafing out. Fifty million years later, plants developed the answer to their third big challenge: how to reproduce without water. The evolution of seeds allowed plants to decrease their ancestral dependency on water for reproduction. Seeds protect the embryo, allowing it to wait until the best moment to develop. It's an arrangement analogous to, as the biologist Lynn Margulis pointed out, a baby born within an egg that only opens when the economy is good.

Timothy Crews is the lead ecologist at the Land Institute. In the drowsy heat of the institute's barn, he explained how waves of life inherit the soil, what is called ecological succession. The ancient microbial mat was the first wave of ecological succession on Earth. In time, subsequent waves of life moved in and took advantage of the groundwork laid by microbes. Imagine, he said, how an island's ecology evolves. A coconut doesn't simply float ashore, germinate, and root in the rock. The first settlers are microbes that can make food from air and rock and water. They jump-start life on sterile land. Microbes convert nonlivable conditions into conditions livable for the rest of us. For billions of years, ocean bacteria and archaea developed ways to convert abiotic, or nonliving, elements like gases into nutritious molecules that living creatures can use to build their cells. When microbial mats made landfall, they brought those skill sets with them.

For example, 90 percent of all nitrogen in the food chain is made accessible to organisms by bacteria that live in soil. Without enough nitrogen, plant leaves turn yellow and weak, and buds grow stunted. In fact, without enough nitrogen, we suffer similar symptoms. Our skin color changes, our muscle mass decreases, and our growth is stunted. The plant gets its nitrogen from bacteria; we get our nitrogen from plants.

Even though Earth's atmosphere is 78 percent nitrogen gas, plants can't use it in this gaseous form. But particular kinds of bacteria are able to change the chemistry of nitrogen from a gaseous form to ammonia, which then converts into nongaseous ammonium in soil. Other kinds of bacteria convert ammonium into nitrate. (Both forms—ammonium and nitrate—are plant-usable, but some plants take up one over the other.) This is called "fixing" nitrogen, as in to alter, not repair. The bacteria that do this provide the plants with nitrogen in a few ways. Some live inside warty nodules on the roots of particular plants such as legumes. Others live in soil, where the nitrogen they fix is absorbed directly by plant root cells. Mycorrhizal fungi also participate when they pick up organic nitrogen, along with other nutrients, from the soil and deliver it to the plant. Whatever plants or microbes don't use is either sequestered in the soil or cycled back into its gaseous form by other bacteria that specialize in reversing these chemical reactions. Ammonium binds to soil, but nitrate doesn't, so what is not used runs off into the ocean where aquatic defixing bacteria take over. In fact, nitrate primarily is fixed on land, but it's primarily defixed at sea. It's a global endeavor, in which nature constantly resets the exchange rates between land, sea, and air.

Other important nutrients that plants need like phosphorus and sulfur are weathered from rock where they are absorbed by plant roots or delivered to the plant by fungi and bacteria. Animals eat the plants, and when the animal or the plant dies, fungi and bacteria decay the corpse. Some of the corpse's nutrients are released back into the environment, some are cycled back into plants again, and some are retained in the bank account that is soil.

It takes at least 1,000 years for the first centimeter of soil to form, for microbes to break the rock apart, for generations of them to live, die, and decay. Rock and air and water launch microbial life, but microbial life launches ecology.

The audience in the pole barn in Kansas grew over the course of the day. I could see tents being raised in the mowed field beside the campus and children venturing farther and farther from the family campsite. The last speaker of the day was John Cobb, a philosopher and environmentalist in his early nineties who spoke about ecological interdependence—that all parts of an ecosystem rely on all other parts. "We are taught to think that the world is composed of separable entities that are separably observable," he said, "but everybody already knows everything is connected." I looked around at the audience. Everyone was looking at each other.

When he finished, we gathered ourselves together. I watched as friends who had spotted one other in the audience clambered over folding chairs to shake hands and hug. Young folks, interns maybe, lugged wood to a bonfire site and hung four black iron pots over a smoldering pit, each one big enough to hold a curled child, full of bison chili that bubbled and erupted in smells bitter and tomatoey and spicy, and a line began to form. Diners balancing plates of food quickly filled an armada of picnic tables plunked around the field, like ships in a harbor of grass. After the sun set, the bonfire was lit and a musical ensemble set up in the barn. A soft-spoken guitarist delivered a long, earnest introduction before strumming the first chords of "Somewhere over the Rainbow" in tones so downy and soothing it had most of the audience members' chins resting on their chests in minutes. A charm of children ran about like furious hummingbirds attacking one another with floppy glow-in-the-dark sticks. If I squinted, I stopped seeing their legs and arms, and only the bits of colored light colliding, circling, separating, colliding again, a spectacle of fireflies at war.

I don't know where Wes Jackson was for most of the weekend. Taking people to and from the local airport, I guess, or talking to someone important like the Pope. But the next day when we went on tours of the test fields, he materialized in the bed of a pickup truck and explained, in his Midwestern twang, that the wild prairie was the tree of life, and that the cultivated fields were the tree of knowledge, and that "we want the tree of knowledge to be subordinate to the tree of life. The tree of life is the standard. And that is perhaps the modern problem. We are all more interested in the tree of knowledge." We walked through an acre or two of wild prairie. As soon as I was surrounded by grass, the sound of clicking insects overwhelmed, muffling my own thoughts in a thick, vibrating blanket. The grass was tall, taller than me. There were countless types, many with small sturdy blossoms, and when I ran my hand through the grass the way girls do in the movies, it was scratchy and rough, a thousand paper cuts waiting to happen. This was grass on steroids.

Actually, it was grass on microbes.

Most life underground takes place in the top 4 inches of soil. Within that small zone, the architecture of soil is built, and microbes are the architects. The biologist Elaine Ingham, who likes to point out that dirt is dead soil, describes soil as a house where the mineral particles are bricks, glued together into blocks by sticky sugars produced by bacteria and fungi.

One particular sticky sugar is glomalin. Nobody paid it much attention until about 20 years ago, but now we know it is pretty important stuff. Glomalin is a carbon-rich protein (a glycoprotein) produced by mycorrhizal fungi that coats the fungal threads like a scuba suit. It's like the silicone you use to caulk your leaky shower. It repels water but also sticks to soil particles. Glomalin lasts in soil for up to 40 years before degrading, even after the fungus has died. The total carbon sequestered in the world's soil is estimated at 1,500 gigatons. A third of that, or 500 gigatons, may be bound up in glomalin at any given time. Just for a little perspective, 1 gigaton is equal to the weight of 12,313 Washington Monuments. Without glomalin sequestering it, there would be a lot more carbon recycling back into the atmosphere as CO_2. And in fact, when CO_2 levels in the atmosphere rise, it stimulates the fungi to make more glomalin, which sequesters more carbon.*

Glomalin is the glue of soil. It helps keep the components of a particle of soil—the minerals and microbes—together, and because of its waterproofing capabilities, it lets water move around the particle without dissolving it. Soil particles stick to adjoining particles to make aggregates, and plant roots and mycorrhizal fungi hold still larger soil aggregates together. One percent of the total fresh water on Earth, in some places up to 9,200 tons of water per acre, is stored in soil structure. Life builds soil structure, and soil structure retains the essentials of life.

The organisms that transform dirt into soil live in a complicated food web just like the organisms in any ecosystem. The bottom line is, everything in life is competing for food, aka carbon, from bacteria on up to us. What the "1 percent" of our species really enjoys is the lion's share of carbon: private jets, imported booze, diamond earrings. But in soil, the real carbon hounds are microbes. Up to 40 percent of a plant's sugars are leached into the soil and consumed by microorganisms. The plant pays soil microbes with sugar, and in exchange, the microbes help with the plant's nutritional and defensive needs. Soil is like Sweden. The plants pay a high tax, but the soil guarantees security.

Plants secrete exudates, a kind of root sweat composed of carbs, sugars,

*That doesn't mean fungi can make enough glomalin to compensate for the 40 billion tons of CO_2 we add to the atmosphere annually.

and proteins that seep into a feeding zone about $\frac{1}{10}$ of an inch around the plant's roots. This area, called the rhizosphere, is the epicenter of the soil food web. It is the microbial marketplace where fungi, bacteria, archaea, and nematodes, a kind of microscopic worm, feed on the plant, feed on the plant exudates, and feed on each other, and in turn are the food of larger critters.

Plants produce these exudates, these delectable seepages, to attract bacteria and mycorrhizal fungi and the nutrients they bring, like nitrogen and phosphorus, and the security they provide by protecting the plant roots from foes. For example, some soil microbes protect plants by emitting substances that can kill other microbes. That's how penicillium mold kills. It produces a chemical that indirectly causes staphylococcus bacteria to explode. Seventy percent of our antibiotics are synthesized from the chemical weapons used by soil microbes. And like our own immune system, which remembers a pathogen (that's the mechanism behind vaccines), fungal and bacterial cells in soils "remember" an enemy organism and specialize in its elimination. These soil microbes are like a mercenary immune system for the plant. It's definitely in the plant's best interest to keep them around on active-duty pay.

Mycorrhizal fungi play an especially big role in plant security. They provide food security by extending the plant's access to water and nutrients and therefore increase the plant's odds of withstanding environmental pressures like drought, and they act as a communication system between plants. One fungus can connect to numerous plants in an ecosystem, thereby connecting the plants to each other. When a bean plant is under attack by aphids, it can transmit a chemical distress call through the mycorrhiza's cablelike network to other bean plants the fungus has colonized, giving them a heads-up there's trouble coming, and maybe buying some time for the plant to shore up defenses.

So while it costs the plant a lot of sugar to pay soil microbes to deliver nutrients like nitrogen and security services like antibiotics and early-warning systems, it's an arrangement that is clearly successful. All plants and animals, from Iowa corn to Anderson Cooper, host microbes that do these two chores: processing nutrients and protecting the mother ship.

Microbes help plants, but they also help each other, and in similar ways. Fungi and bacteria cooperate by trading food, providing succor, and driving away adversaries. They also help each other kick-start behaviors. For example, some bacteria produce chemicals that make the fungus *Candida albicans* grow

(*C. albicans* is responsible for our yeast infections), and the fungus produces a chemical that helps *Pseudomonas aeruginosa*, a bacterium that causes disease in plants and animals, become virulent. The bacterium *Klebsiella aerogenes,* which is ubiquitous in everything from flowerpots to your mouth, produces a precursor chemical that allows the fungus *Cryptococcus neoformans** to make melanin, the same pigment that is in our skin, which provides it (like us) with UV protection. All the microbes in soil—and all their different by-products— are interacting with each other, even communicating. Bacteria and fungi may actually talk to each other using a language of chemical fragrances called ter-penes. They don't have noses, but they can produce and detect and respond to each other's terpenes. They speak in perfumes.

But bacteria and fungi aren't just involved in their own complicated trans-actions with plants and with each other. They are implicated in the lives of the larger organisms, as well. Hungry protists are generally 10 to 100 times larger than bacteria, about the ratio of a watermelon to a pea. They can eat 10,000 bacteria a day and will eat each other if bacteria cannot be found, a gruesome sort of population control. Protists have variable hunting methods. For exam-ple, amoebae engulf bacteria, and ciliates move about in the watery interstices of soil particles like little hairy rowboats, beating water to bring bacteria into their mouth region.

Microscopic nematode worms are also bacteria gobblers (and so compete with protists). These are tiny animals, only 5 to 100 times larger than bacteria, though they can be very big, too. The largest, called *Placentonema gigantissima,* is about the size of your average garden hose (it parasitizes the placenta of the sperm whale). There are up to a million species of these tiny worms, and while some are root-feeders, others eat fungi and bacteria—up to 5,000 bacteria a minute. They help keep microbes that could attack plants at bay, and in "chew-ing up all sorts of proteins," said the microbiologist Philip Strandwitz (he means other critters), "they squirt out juicy nitrogen that is plant-ready for con-sumption." But just to reinforce the idea that soil is a food web, not a food chain, some fungi turn around and eat nematodes. Carnivorous fungi set

C. neoformans is often found on bird feces. It causes a rare but dangerous illness that almost killed Bob Dylan and is the reason why there used to be "Don't feed the pigeons" signs around cities. You do not want to breathe those dried droppings.

hyphal snares for root-eating nematodes, which squirm in, but not out. There are nematode species that eat protists, algae, little insects like weevils, and other nematodes.* Nematodes constitute 80 percent of all individual animals on Earth; in fact, the "father of nematology" Nathan Cobb observed, "If all the matter in the universe except the nematodes were swept away, our world would still be dimly recognizable . . . by a film of nematodes."†

Insects are the microshredders of soil. They chew up plant debris into pieces that fungi and bacteria can handle, much as we chew our food to make the bits small enough for our gut microbes to manage (and microbes in the guts of insects digest the plant matter for them). Insects like mites and springtails are responsible for shredding 30 percent of the debris deposited on the temperate forest floor. You might have seen springtails before. They're that cloud of little black specks that pop out when you move a rotten log. Sometimes called soil fleas, they don't bite. But they do jump really high, the equivalent of a human jumping over the Eiffel Tower. Termites and ants are shredders, too. They break down plant matter, and their endless tunneling mixes and aerates soil—harvester ants in Florida have been found to mix a ton of soil per hectare a year—to the benefit of all oxygen-breathing critters. You want ants and termites and springtails in your garden.

But it's the earthworms that are the heavy terralifters.

An acre of garden soil can contain over one million earthworms. I like to think of them as free-living intestines. They even look like intestine. As Charles

---

*And some cause disease like hookworm infection. Hookworms infect the human body through the feet and float up the bloodstream to the guts, where they attach to your intestinal wall to feed. Mostly they are considered parasites, and certainly an overinfestation can cause anemia, but there are those who believe underinfestation may be the culprit behind allergies and other autoimmune disorders. Some folks believe that humans and helminths (hookworms and whipworms) have evolved together; we give the helminth food and safe haven, and the helminth gives our immune system a positive regulatory advantage. The idea had enough oomph to warrant clinical trials, but researchers haven't established whether or not worms help with immune problems like asthma.

†Tardigrades, 530-million-year-old aquatic micro-animals, eat up to 60 nematodes a day, as well as bacteria and plants. They are about as big as the point of a very fine lead pencil, but they've got many of the working parts of a bigger guy: esophagus, stomach, intestine, anus, muscles, a brain atop a nervous system. They are incredibly rugged, able to take extreme heat, pressure, and suffocating gases. They can dry up and stay in a mummy state for up to 100 years, even in the vacuum of space, and then be revived by a drop of water. These qualities have spawned a minor industry in "tough as a tardigrade" swag. Tardigrades even have a protein that confers the ability to tolerate doses of radiation that would kill most animals. This protein has been experimentally transferred to human cells, where it makes them resistant, too.

Darwin wrote in his 1881 book *The Formation of Vegetable Mould, through the Action of Worms*, every particle of soil has been through an earthworm at least once. Their burrowing aerates soil and improves its water-holding capacity by keeping it from becoming compacted, and their burrows are rich in bacterial and fungal growth, thanks to their delicious slimy excretions. Earthworms are shredders. Leaves that would take 1 to 2 years to decay are composted in 3 months by worms. They consume soil in search of bacteria, fungi, nematodes, protists, and plant matter. Their gizzards break down food, and, like all animals, bacteria living in their GI tracts digest it. Nutrients are absorbed in the worm's bloodstream, and the rest is eliminated as plant-ready, nutrient-rich castings, fertilizer gold for farmers.

And earthworms attract birds, whose droppings are rich in nitrogen. Plants need the nitrogen, which allows them to conduct photosynthesis, which terrestrializes carbon, much of which percolates into the soil for the benefit of fungi and bacteria, which feed the earthworms. It's like soil digests life on Earth. Indeed, soil is the planet's gut microbiome.

Because there was no relief from the studying, I had to bring my 7-pound biology book with me to Kansas, which was hassle enough, but as I passed through Wichita's calm little airport on the way home, I got pulled off the security line because of it.

"It's just a biology book," I said. But the TSA agent was unimpressed.

"We have to check it anyway. We don't see these very often."

What? He didn't often see a thousand-page, multipound book being willfully lugged across the country? In retrospect, I suppose it was suspicious. I watched as he riffled through the pages, and then I got it. He was looking for a secret compartment. A place where I might hide my stash. I did, actually, have a small bag of prairie dirt in my purse as a souvenir, a small bag of a vast but tiny food web. But he didn't ask me about it. I suppose that was more oddball than contraband.

Back in class, we had a few more lectures that blazed through the nervous system and the sensory systems—the only resonating fact still buzzing in my head was fish have taste buds that run the length of their bodies—and animal digestion, the juices of the pancreatic system, the neural and endocrine reflexes that stimulate gallbladder contraction. At times, I felt like I could have been studying cosmology, only without the math. Doesn't the "reflexive wave of peristalsis" sound like it could have been about space? But besides a bit of attention on the 500 milliliters of gas the bacteria in our colon produce as a result of fermenting fiber (more if you eat a lot of beans) and the 50-gallon fermentation vat that is the cow gut, we would have to wait until next semester to learn about our gut microbiome. Why? Because the gut microbiome is not covered under the section Animal Form and Function. It's covered under Community Ecology.

During these slogging days, I would periodically log on to Piazza, Columbia's chemistry forum, just to see how the class I'd bailed was faring. "Why is nitrogen trichloride (NCl3) polar? Aren't all of the bonds neutral when the central atom is drawn with a lone pair?" asked one student.

It helped remind me just how much worse it all could be.

Shortly before the next test was scheduled, I happened on some of my son's ADHD medication that was left over from a rocky sophomore year in high school. The drug, Focalin, is dexmethylphenidate hydrochloride, a central nervous system stimulant. It's similar to Adderall and a popular study drug. One out of every six students in American universities used Focalin or one of the other half dozen ADHD drugs in 2015. And that number is likely low. Stimulants just seem to come with the student territory. My fellow students drink Red Bull all the time, and there is always a line at the campus Starbucks, snaking through the store like a Nigerian voting queue.

I waffled for quite a few days about trying it, during which time I felt guilty and upset with myself for not having taken the drug before I served it to my child. Six and a half million American children ages 4 to 17 are diagnosed with ADHD, and ADHD online support groups are full of worried parents dealing with young medicated children who lose weight, can't sleep, or develop Tourette's syndrome, pulling out their eyelashes and picking at their skin. Some private middle and high schools, here in New York anyway, are notorious for pushing ADHD drugs on underperforming students, and public schools seeking to reap score-based rewards have used ADHD diagnoses to exempt low-performing students from the school's overall achievement rating. The rise in ADHD drugs is due to the selling of the disorder, explained Alan Schwarz in his book *ADHD*

Nation, by the companies that make these drugs, by the doctors that diagnose the disorder, and by social pressures that demand academic distinction.

I figured I'd take it during my next test. It did occur to me that taking the drug was a form of cheating. Not like with athletic performance drugs, which are against the rules, but because they reinforce certain advantages. These drugs aren't cheap. Well, actually, the drugs aren't expensive, but getting the prescriptions can be. We paid $500 for our son's appointment to see a pediatric pharmacologist, and insurance didn't cover it. A student that can't afford ADHD drugs or is otherwise indisposed to using them is at a competitive disadvantage. And there's no easy solution. Handing out enhancement drugs before a test to level the playing field or banning them (and the related enforcement problems) are problematic choices. Anyway, they are pretty easy to buy on campus. A 30-milligram dose of Adderall from a student dealer costs about $5, less than a Pumpkin Spice Latte. In a 2010 study of 81 college students with ADHD, 62 percent diverted the medication to someone without a prescription.

Performance enhancers are used all the time in our wider society. Take Viagra, or microdosing hallucinogens, which reputedly allow for maximum creativity, minimum ego. The authors of a controversial opinion essay published in the journal *Nature* in 2008 suggested mentally competent adults should be able to engage in cognitive enhancement drugs. "A modest degree of memory enhancement is possible with the ADHD medications . . . which raises levels of acetylcholine in the brain. . . . It is too early to know whether any of these new drugs will be proven safe and effective, but if one is it will surely be sought by healthy middle-aged and elderly people contending with normal age-related memory decline, as well as by people of all ages preparing for academic or licensure examinations."

I seemed to fit the bill, so I took a short-acting dose on the subway train up on the day of the test. My heart raced! My ideas were rolling! I felt hyperfocused! I was feeling super positive when I sat down with the test in front of me. I was raring to go! That's the good news.

The bad news is when you take drugs that help you focus, you don't necessarily get to choose what you focus on. I spent the entire hour with half my mind running through the lyrics of the New York Mycological Society's Hymn to Mycology, which celebrates the degrading powers of fungi and is famous in our club for turning the stomachs of unsuspecting listeners.

This test was, once again, 15 pages long, with horrible questions like "Are protonephridia found in open circulatory systems just like Malpighian tubules?" (I got that one wrong) and "Do Malpighian tubules excrete wastes through spiracles?" (I got that one wrong, too). By suppertime the night of the test, I was back to my normal just-barely-staying-awake-past-9:30 self again, and I returned to a state of low-grade anxiety as I checked my notes against the questions I was afraid I got wrong. The order of events of a sarcomere: Does depolarization move down through the T tubules before or after the voltage-gated calcium ion channels of the sarcoplasmic reticulum open? I couldn't even remember how I answered that one, but I assumed I got it wrong (I did).

I also looked up the lyrics of our infamous hymn.

> Where root and stump lie mouldering,
> 'neath leaden dripping skies,
> There, there shall we foregather
> As unwholesome vapors rise.
> Deep, deep in the murky shadow,
> There where the slime mould creeps.
> With joy the stout mycologist
> His pallid harvest reaps.
> No cloud of noxious insects,
> No landlord's squamose heart
> Can stay our dedication
> To the mycologic art;
> As tramping on into the gloom
> Right lustily we raise
> From every loyal gullet
> An anthem in thy praise:
> Mycology! Priapic muse!
> Great Goddess of decay!

I love that hymn. Not just because it is really unique, but also because I admire creatures that return the elements of life back to the pool of opportunity. I think detritivores, which feed on dead and decomposing things, are optimistic, though they are definitely not everybody's bag. I know folks who won't touch a mushroom out of revulsion for its connection to death. But decomposition is the resurrection part of the cycle of life, and it's an important role played by fungi. All living things die, and when they do, their carbon and other nutrients are dismantled into the building blocks of life again.

But keep in mind, we decompose according to our scale. What goes up in complexity must disassemble the same way. Top predators are broken down by top detritivores. In the Tibetan sky burial, a naked human corpse is transported to the top of a mountain, the flesh flayed and left exposed to be eaten by large carrion birds. When the carrion bird dies, its carcass is shred by carnivorous varmints and arthropods like beetles (of which there are 350,000 species, so many that the witty British scientist J.B.S. Haldane pointed out the Creator seemed inordinately fond of them). Fungi and bacteria break down the remaining pieces into their raw ingredients, like nitrogen, hydrogen, sulfide, and phosphorus, and other bacteria help recycle these molecules back into the system. Fungi most certainly absorb the carbon of the dead, but they don't do the job alone.

That's why the artist Jae Rhim Lee's Infinity Burial Suit, which is a pajama threaded with fungal spores bred to decompose bodies, while a cool idea that reminds us we are all participants in the cycle of life, doesn't really do what it claims to do. The suit costs $1,500, as does the Infinity Burial Shroud (there's an "add to cart" button, which I think is a much more sensitive choice than "check out"), but they won't buy you a speedier decomposition. Fungi are actually pretty bad at decomposing corpses. We bury our dead deep, and fungi are aerobic. They need air. The truth is, not much will happen to a corpse or anything organic quickly if buried 6 feet under the soil's surface, because soil microbial diversity drops off the deeper you go.

If it's efficient land-based breakdown you are after, according to Jennifer DeBruyn and her team at the Department of Biosystems Engineering and Soil Science at the University of Tennessee, it's probably best to bury grandpa under a pile of woodchips, which has lots of little air pockets to keep the decomposers alive and provides additional carbon for the microbes that offsets all the nitrogen in him.

It takes a corpse months to degrade. When an animal dies, oxygen stops going in and CO_2 stops going out. The acidity level increases and the cells collapse, releasing enzymes that break down surrounding tissues. "The enzymes that built us," wrote William Bryant Logan in *Dirt*, "now undo us."

Without fresh oxygen coming in, our internal bacterial communities change and our decomposers dominate—yes, we carry around the laborers of our own decomposition. "There is life after death," said Dr. DeBruyn in the magazine *Science News*, "and it is mainly microbial." The decomposers break down the proteins in our cells and produce by-products like methane and hydrogen sulfide gas that bloat and rupture the body. Bacteria also produce the nasty smells of the dead that attract insects. Insects disassemble the bulk of the body mass, and the low pH of the enzyme soup attracts fungal decomposers. If there are any pathogens on the corpse, microbes in the soil kill them, or they die of exposure or lack of food, which is why graveyards aren't hotbeds of disease.

If the mineral parts of soil provide the stage for the drama of microbial life, it's the dead that supports the action. Life in soil springs from a dense nutrient-rich mantle of dead plant and animal residue and decaying microbes called humus. Humus is like the currency of soil. Living organisms withdraw carbon stored in dead cells, and what's not spent builds up over time like money in the bank. Ultimately, the richness of soil is based on its savings of humus. And Earth's endowment of humus determines the livability of most terrestrial life, because we depend on what grows in the soil.

The word human is derived from the Latin *humus* for "earth." In Hebrew, Adam is related to *adamah*, which means "soil," and Eve to *hava*, which means "living." Soil, humus, human, living: We are connected both figuratively and literally to the microbes under our feet.

Soil without Microbes

I don't know why I didn't understand this before, but my biology class was taught in relays, and it seemed like just as I was getting used to Professor Naeem and his encyclopedic study sheets, his portion of the class was over. His goodbye email hoped we found his segment "challenging, thought provoking, but fair—no surprises," except for the biggest surprise of all, that he was gone, and another professor was stepping into his loafers. I ended up getting about the same grade on my second test, so I was actually feeling like I had reached a kind of academic equilibrium, but once Professor Naeem was gone, my newfound confidence went with him. What's more, his replacement seemed significantly less interested in coddling us by calling us environmental biologists and grading on a curve.

Professor Dustin Rubenstein was a young guy, relative to me, meaty and vigorous. His lab studies how animal social evolution responds to environmental pressures, focusing mainly on birds in Africa and the inch-long snapping shrimp, a creature whose claw clicks reach decibels of 190 (your eardrums rupture at 160), about as loud as the call of a 40-foot sperm whale. But what he taught us in Biology 101 was genetics.

The section covered the different ways genes are stored and sliced and diced and copied and analyzed, including the polymerase chain reaction, or PCR, a cheap and easy way of cloning DNA from a miniscule sample, like a single skin cell. One of the people who figured out how to do this was the surfer biochemist Kary Mullis. Natural DNA replication happens inside a living cell.

PCR does that job *outside* the cell. To prep the DNA for copying, it has to be exposed to high heat, and high heat destroys an important enzyme that is necessary to complete the copying process. What Mullis discovered was if he used an enzyme from a heat-loving bacterium found in a hot spring in Yellowstone National Park, he could bring the DNA to the high temperature necessary for the reaction to proceed. Mullis got a Nobel Prize for the technique, which enabled scientists to do massive DNA sequencing like the Earth Microbiome Project and the Human Genome Project. It essentially started the genome revolution. "If someone gives me an idea that wins a Nobel," said Professor Rubenstein, "I'll give you an A."

One night I went to PCR & Pizza, a free evening at Genspace in Brooklyn. Genspace is part of the nationwide garage lab movement, also known as DIYbio, where amateur and professional scientists can conduct biologic experiments in a nontraditional lab setting—it's also sometimes called biohacking—usually slapdash spaces with jerry-rigged equipment bought on eBay, like centrifuges made from commercial eggbeaters. It's punk biology.

I had to wait on the street outside Genspace's building for someone to come downstairs to let me in. There was nothing so official as an intercom, and it took me a while to figure out the birdsong I was hearing was a recorded loop emanating from somewhere above the door. The building is a warehouse for all kinds of interesting businesses, like interactive design groups, and engineers that make electric skateboards, as well as a you-name-it-he's-got-it storage facility for the owner, a Santa Claus-meets-Jerry Garcia-like idealist named Al Attara. Genspace doesn't look like a lab in a movie. It's lined with repurposed stainless steel restaurant counters piled high with boxes of supplies like petri dishes and rubber gloves, electric wires and Christmas lights hanging overhead like Spanish moss, trash cans stuffed with pizza boxes, an array of machines jiggling and buzzing with mysterious purpose. When I sat down on a well-worn couch, a few moths puffed out.

Ellen Jorgensen is one of the founders of the lab. Brown-haired and assured, I watched her TED talk, "Biohacking—You Can Do It, Too," where she explained what biohacking is (amateurs and rogue professionals doing science) and what it does (democratizes science) and how it can be used. Personal biotech allows you to check if your sushi really is tuna or to explore your ancestry.

One biohacker in Germany wanted to know whose dog was leaving a little present on his street every day. He threw tennis balls to all the dogs in the neighborhood, analyzed the saliva, matched it to the poop, and confronted the dog owner.

PCR & Pizza wasn't really about PCR—cloning genes—it was more about how you isolate genes from a sample, with frequent pauses to sip beer. Dr. Jorgensen had our group harvest cells from inside our cheeks with a cotton swab ("I twirl it for 5 seconds," she said) and then combine our cells with a saline solution in a plastic tube with a top, each with our initials. The cells were spun in a centrifuge to separate them from the saliva, where they fell to the bottom of the tube like a pellet. Then we added a drop of tiny beads and spun the tubes again, which broke up the cells and spilled out the cell contents. Finally, the mixture was boiled to denature the proteins. The proteins fell out, and the double helix structure of DNA temporarily fell apart, too, "but DNA is hardy," said Dr. Jorgensen. "It snaps right back together." It was also dense and didn't drip when I turned the tube upside down. My entire genome was reduced to a clear sticky droplet in the bottom of a plastic tube, which I promptly lost in the void of my purse.

A couple of weeks studying genes in biology class had prepared me for some of the things we learned at Genspace, because when Dr. Jorgensen mentioned that had we enough time, we would have looked for a particular gene called *CCR5* that confers HIV resistance to double mutants, I understood that to mean you probably wouldn't get AIDS if you inherited the mutant gene from both parents. In class we'd learned about alleles, alternative forms of a gene that arise by mutation, some of which are dominant and others recessive. We studied inheritance, like baldness and albinism and hemophilia, and ultimately the work of Gregor Mendel, a 19th century Austrian abbot who (though others proceeded him in the work) showed how observable traits, like whether a pea flower was purple or white, were distributed among offspring and how those traits' distribution could be anticipated. Before Mendel, inheritance was viewed as blood-borne traits (hence the term *bloodlines*) that were blended in the next generation. But since Mendel, we've known how to figure out what the chances are your child will have brown eyes.

I actually hated our inheritance studies, especially the game of determining probabilities. We were taught how to use a Punnett square, a tabulation

diagram that predicts the chances of one's offspring having a particular genotype, and I was fine when the diagram was really simple, what's called a monohybrid cross, which might ask, what's the chance a child will have brown eyes
like Mom or green eyes like Dad? But anything more complicated—like what's
the chance a child will have Mom's brown eyes *and* Dad's height?—and I mentally shut down.

With genetic predictions, I bumped up against the insurmountable. I could
not get my head around probability, the branch of mathematics that deals with
calculating the likelihood of something happening. I reviewed the chapters on
true-breeding, reciprocal crosses, monohybrid crosses, and dihybrid crosses
over and over. They were written in English, but they might as well have been
written in Chinese for all I understood. I just couldn't grasp it. It was like trying
to grab something on a shelf that I was just too short to reach.

Professor Rubenstein emailed us a few pages of inheritance practice problems that I dutifully answered, but then when I checked the answer key, I found
out my results were completely wrong. So wrong, in fact, that I didn't even
know *why* they were wrong. In the midst of all this, one afternoon after class I
caught up with Professor Rubenstein as he walked back to his office and tried
to explain my predicament, but he told me to just go online and find more
problems.

Which I did. One weekend I spent 10 hours straight working on online
inheritance problems, but even the ones targeted to 10th graders stumped me. (I
discovered that when looking up anything in biology for the first time on the
internet, I was best off if I used the keyword "high school," as in "reciprocal
crosses high school.") When I'd read, "If a woman is homozygous normal and
her husband is heterozygous for a genetically inherited recessive disease, and
they decide to become parents, what is the probability that they will have a
healthy child?" my mind felt numb, like it does after learning bad news. I did
hundreds of these problems, filling pages with bogus calculations. I watched
hours of YouTube videos, falling briefly for Nikolay the Russian geneticist's
accent: "So now when we beeld Punnett square once again we predickt genotype
of progenies." But I could not correctly answer any question the first time. Ever.

I considered finishing the semester and then dropping out. Because this
was getting too hard, because I was spending so much time studying, because
my personal life was getting chaotic. I was behind in everything: the household

accounting, my work commitments, visiting my parents. Weeks went by without clean sheets or toothpaste. A friend came to stay for a few days, and I only showed him how to work our rather daunting espresso machine once, just once, and thereafter when he wandered into my office with the coffee filter in his hand and a confused expression on his face, I told him to go to a fucking Starbucks. This was different from just working hard. Since we started studying genetics, I was working really hard to no avail. Working hard and accomplishing very little is humbling. Working hard and still being a total loser just seems idiotic after a while. I wanted to quit.*

Maybe some people just aren't math people. I only tip waiters 20 or 25 percent because I get flustered if I have to figure out a more complicated number. Math is a genetic ability to some degree, but it may be that my own sense of doom is what really tanks my math abilities. According to various studies, some students think their abilities are nonmalleable, that regardless of effort you can only do what you can do. Others believe intelligence is malleable, and you can increase it with effort. I think both are true. I do believe ability is malleable, otherwise I wouldn't have tried to learn biology at all and I would've stuck to inventing new panna cotta recipes, but I am also quite sure I can't calculate how fast a snowball will melt if its radius is x inches. Not in this lifetime.

But the class wasn't all odious Punnett squares. We also learned how bacteria help introduce new genes into other organisms like plants. Bacteria have two loads of DNA in their cells. One is a dominant tangle that codes for the normal functioning of the cell. The other is the plasmid, a circular rope of DNA with fewer genes. Bacteria share genes by sharing copies of their plasmid (by means of their penislike pilus. Ah! Now you remember). Plasmids are not necessary for a bacterium to live, but they provide survival advantages. Antibiotic resistance, for example, happens when a bacterium containing a plasmid with

---

*According to *America's College Drop-Out Epidemic*, a report from 2014, 40 percent of full-time 4-year college students fail to earn a bachelor's degree within 6 years. The reasons run the gamut from family problems and loneliness to academic struggles to a lack of money. Columbia manages to retain 99 percent of its freshmen, significantly better than the national average, despite the tuition. My biology class cost $7,100 per semester. I actually hid the cost from my husband for a couple of months. It was like buying an expensive dress that you stash in the back of the closet until your guilt settles down. If I wanted to start over as a freshman, it would cost me $53,000 per year, not including a slew of fees, housing, transportation, and extortionately expensive textbooks. My annual tuition, including room and board, at Barnard in 1982 was $12,700, which my parents paid.

an antibiotic-resistant gene shares it with a recipient bacterium that didn't have the gene, and boom! Now both bacteria are antibiotic resistant.

To get a new gene into a plant, biotechnologists start with the plasmid of a bacterium that infects plants. They remove the plasmid and cut it open at a particular gene site, kind of like cutting open a jacket at the seam, and insert a piece of foreign DNA, say, a gene that allows plants to tolerate an herbicide, like putting an extra panel into the jacket to accommodate your growing girth. Then the plasmid is put back into the bacterium. The bacterium, now carrying the new gene, is allowed to infect the cells of a soybean plant, for example. From those infected soybean cells, new plants can be cultivated containing the gene.

That's what a GMO, a genetically modified organism, is. The vast majority of research on GM crops suggests they are safe to eat. That doesn't mean problems won't be discovered in the future. Genomes react to the presence of alien genes, and inserted genes can be transformed within the genome over time. But what *is* certain is that agricultural inputs, like the herbicides GM crops have been bred to tolerate, undermine the diversity of the soil microbiota. And that's definitely a problem.

When a farmer speaks of rich soil, she is talking about microbially rich soil. When you smell the sweet scent of fresh soil, what you smell is geosmin, an organic compound produced by the bacteria *Streptomyces*. That most profound of perfumes, the smell of promise and fertility and all that is good and eternal about our planet, is the by-product of a microbe.

The value of a piece of farmland, besides access to water, is based on fertility, and fertility can be defined as the land's population of microbes and the jobs they do, like harvesting and delivering phosphorus to plants. A lot of modern farming replaces those functions with synthetic solutions like industrial fertilizer, in effect, bypassing the roles of certain microbes. It's like we've automated soil and the microbes are out of a job.

Soil has always been a measure of wealth, whether that measure is a French country girl's dowry based on the amount of manure produced on the family farm or how an Iowan farmer is taxed based on the productivity of his land. The wealth of soil is a reflection of its diverse microbial content, and the poverty of soil is a reflection of the lack of diversity. How to grow food without permanently depleting or destroying the soil is an age-old problem. In *Topsoil and Civilization*, a manifesto on soil preservation from the 1950s, the authors argue that a progressive civilization rarely lasted in one locality for more than 30 to 70 generations, or 800 to 2,000 years. Almost all succumbed to a failure of their farmland to produce an adequate supply of food due to overgrazing, farming hillsides, and monocropping nutrient-greedy plants like cotton. The ancient world is rife with examples of man-made soil degradation, salinity, and erosion. Crete's farmland built the Minoan civilization; now, it's a stony waste. The topsoil of the Attica plains in Greece once fed a great civilization, but now it's only a few inches deep. It happened in the USA, too. It only took 30 years prior to the Dust Bowl of the 1930s for American farmers in the southwestern Great Plains to torpedo their land with soil-depleting farming practices. Those practices were founded in the ideas of an 18th century Englishman named Jethro Tull.

Not to be confused with Jethro Tull of *Aqualung* fame, which was a huge hit when I was in middle school. My friends and I lapped up the big riff-heavy songs like chocolate milk. We had no idea that the band was named for an English agriculturist whose contributions to farming set the stage for industrial agriculture in this country and, in some cases, caused microbial genocide. Jethro Tull, a well-to-do farmer from Berkshire, England, argued that a healthy field was a weedless field and that the addition of manure, with its payload of grass seeds of all sorts, introduced weeds. He argued for deep and constant tillage instead. "Tillage," wrote Tull in his book *Horse-Hoeing Husbandry*, "is manure." He believed that earth was the sole food of plants and the smaller the particles of earth, the easier it would be for the plants to feed.*

*Eighteenth century agricultural reformers advocated the tillage system, but it was not a good fit for all landscapes. It was tillage as much as potato blight that caused the Irish famine in the 1800s. In traditional Irish farming, potatoes were planted in raised beds, and water drained away from the plants into furrows. But by 1834, Irish farmers converted to the Tull method, to their despair. Raised beds discouraged the potato pathogen *Phytophthora infestans* (a funguslike protist) by drying out wet soil, but the deeply tilled, flat fields lacked this advantage, and so, in combination with planting a single variety of potato, wrote Charles Mann in *1493*, "on a terrain shaped for technology, not biology," the pathogen took off, causing a famine that killed a million people and sent waves of early Irish immigrants to the USA as economic refugees.

Tillage actually does produce a boon crop, but not because plows cut up sod into bite-size pieces. When virgin soil is plowed, it breaks apart the microbe-built structure of soil that was managing nutrients in a slow-release pattern. All of a sudden, a flood of nitrogen, phosphorus, and sulfur necessary for plant nutrition and microbial growth becomes available and a feeding frenzy ensues. For a couple of decades, the crop gets an incredible boost, but when the nitrogen and other nutrients stored in soil and crushed fungi are used up, the microbial diversity drops and the crops become more susceptible to disease and nutrient deficiency. When the soil food web is undermined year after year by tilling, the community of microbes never has a chance to develop beyond a certain degree of functionality. It's like letting a civilization get to the point of producing bronze and then bombing it back to the Stone Age again.

Tillage is a boom-and-bust enterprise that ultimately leads to soil degradation. Dead plant matter isn't left on top of the soil long enough to feed microbes. Fungal networks are destroyed and can no longer retain water, sequester carbon, or deliver nutrients like phosphorus to plants. Arthropods like earthworms are squashed. (Worm segments *can* grow back, but never more than they lost, and definitely not if they lose their head.) Tillage doesn't destroy microbial abundance—there is and always will be some type of bacteria that can grow anywhere—it destroys microbial diversity and the complexity of the soil society. And when that complexity is reduced, so too are the number of services it can provide. If an uncommon pest attacks a tilled field of corn, for example, there may not be the microbes on hand to help fight it off. And if there is a drought, the soil, no longer bound by microbes into particles and no longer capable of retaining water, dries up, and as the Dust Bowl illustrated, it will just blow away. And if soil is gone, "the loss is irrevocable," said Land Institute cofounder Wes Jackson. "Currently US soil erodes at a nonrenewable rate. Soil is as much a nonrenewable resource as oil."

Tilling, thankfully, is on its way out. In some cases, farmers are using low-till (some tilling) or no-till, where planting is achieved by drilling just the hole for the seed in the ground. But what farmers have done to maintain the high yields of tilling is replace the microbial roles in soil with synthetic inputs, to, in the microbiologist Krista McGuire's words, "force-feed" the soil. This practice is common on the family farms of Iowa. When I think of a family farm, I picture a few generations living together, raising organic crops and animals. But those kinds of farms exist in what Wes Jackson calls "donuts" around cities. A family farm like Ron Heck's in Perry, Iowa, is closer to the actual model. He farms 4,000 acres of corn and soybean alternatively. With one other person.

I met Ron through his uncle, Mike Heck, a friend from Colorado. Mike and his wife, Barb, are the kind of people who, when you run into them in the café, explain they've just come back from Mongolia. Two *months* in Mongolia. When I first reached out to Ron, I sent him a copy of my book about mushrooms.

"I read your book this week," he wrote, "and enjoyed it. Most of it, anyway." That set the tone for our acquaintance. We were going to start with me on probation. But I understood why. Ron's farming methods produce vast quantities of corn, but they come at an environmental cost. "This farm produces a basic diet for 100,000 people," he told me as we stood on top of an eight-story corn silo and looked over flat fields stretching away into the haze, continuous as a ballroom carpet, a sight best articulated with a sigh. "As Iowan farmers, we provide affordable, safe food. When we don't grow an acre of food here, they plow up an acre of forest in Brazil."

Ron's two-man production is possible because of mechanization, nitrogen fertilizers, pesticides, herbicides, and biotech seeds that tolerate weed killers. "I guess you'll hear chemicals destroy microbes and no one cares," said Ron, "but truth is the state of the science does not tell us how to economically take care of the microbes in the soil." A farmer like Ron can't be assured the kind of profits that allow a family to put a bunch of kids through college unless he bumps up his productivity with synthetic fertilizers. But give him a natural microbial inoculant that delivers the same financial results as nitrogen fertilizer and he'll use it.

"If I want to treat my soil better, I can go to a crop rotation," he admitted. That's the practice of growing different crops on a given field in a sequence of seasons. Instead of the same crop sucking up the same nutrients, crop rotation allows for a diversity of nutrients to be reestablished in soils. "My soil will get healthier, but by not producing food, you could say 100,000 people will starve. So in this sense, it is unethical not to hurtle toward our own destruction through environmental degradation." That's a moral argument I've heard from other farmers, as well. Feeding people is a mantra that conveniently goes hand in hand with an economic argument. If you take a year off to improve your soil with a cover crop like vetch, even if the corn is more abundant the following year, you still make less money than if you just kept growing corn continuously in the same fertilizer-subsidized soil. For Ron, that boils down to a choice between growing food and growing microbes. "I'd like to feed my microbes," he said as we clambered down the silo ladder, reentering the sound zone of crickets, "but they've got to feed me back."

The problem is many crops like corn don't fix nitrogen on their own. If you want to grow corn, you have two choices. You can rotate the corn crop with a plant that has a symbiotic relationship with nitrogen-fixing bacteria, like legumes, to get plant-ready nitrogen into the soil. The Ancient Egyptians did this by actually doing a kind of soil fecal transplant,* where shovelfuls of soil from fields enriched by nitrogen-fixing fava crops were transplanted to fields growing non-nitrogen-fixing crops. Or you can add synthetic nitrogen.

How much nitrogen there is in the soil limits the amount of growth a plant can accomplish. The hybrid corn seed that Ron uses is greedy for nitrogen, so he adds synthetic nitrogen fertilizer to his fields. Synthetic nitrogen is made by using hydrogen, heat, and pressure to convert atmospheric nitrogen into ammonia, the chemical precursor of synthetic nitrites and nitrates. The German scientists who developed the technology, Fritz Haber and Carl Bosch, both won the Nobel Prize, but not without controversy. Haber developed chlorine gas and Bosch developed nitrogen explosives, both of which were widely used with lethal effect during World War I.

The United States used the Haber–Bosch process to make munitions like TNT during World War II, but after the war, the industry shifted to fertilizer production, and today 1.3 percent of energy used in the USA is employed in the production of nitrogen fertilizer. The industry that produced yesterday's bombs is producing today's food, a fact that elicited boos when introduced in our biology class. But the use of synthetic nitrogen fertilizer is one of the drivers of the postwar population boom. "Without Haber–Bosch," said Wes Jackson, "40 percent of us wouldn't be here now."

*A fecal transplant is the transfer of human stool from a healthy donor to the gastrointestinal tract of an unhealthy recipient in order to treat certain diseases by introducing beneficial gut bacteria. More on that, later.

Nitrogen isn't the only ingredient in synthetic fertilizer. Phosphorus and sulfur are in it, too. In nature, fungi supply phosphorus and sulfur to the plant in exchange for sugars. But when synthetic phosphorus and sulfur are added, the plant doesn't need the fungi anymore, and the fungi withdraw and the density of fungal networks declines. Farmers like this because corn that depends on mycorrhizal fungi gives away more of its sugar than corn that has been artificially fertilized. But less fungi means less glomalin, and without glomalin, soil is less secure, holds less water, and is more likely to blow away.

The application of chemical fertilizers—typically 100 to 200 pounds per acre of land—increases a farmer's yield by almost 100 percent and allows for continuous corn cropping. The same increase occurs when corn is planted immediately after an established legume, but then yields are biyearly. Out of 150 or so pounds of fertilizer, about 20 percent of synthetic nitrogen (either nitrate or ammonium that is then converted to nitrate by soil microbes) is lost to leaching. When it rains, the nitrogen not taken up by crops flows into first local, then regional, and eventually national water supply systems where it fertilizes the growth of phytoplankton. When the phytoplankton die, they are eaten by bacteria that grow to such huge numbers they use all the oxygen in the water, causing fish die-offs.

Agricultural runoff is one of the main contributors to the hypoxia, or suffocation, of the Gulf of Mexico. The "dead zone" in the Gulf, adjacent to the Mississippi River, is the size of Connecticut and Rhode Island combined. It's the second largest in the world after the Baltic Sea. (There are over 400 dead zones worldwide.) According to a review paper in the journal *Science*, "the key to reducing dead zones will be to keep fertilizers on the land and out of the sea. For agricultural systems in general, methods need to be developed that close the nutrient cycle from soil to crop and back to agricultural soil." A great deal of the nitrogen that's killing off aquatic life in the Gulf is coming from Iowa.

I asked Jim Gillespie, Iowa's Division of Soil Conservation & Water Quality's amiable director, what Iowa was doing to stop the nutrient hemorrhage. "We try and give farmers alternative practices" through cost-share programs, like creating wetlands, an edge-of-field practice that removes nitrogen from the water, and providing information that helps farmers meet conservation and environmental objectives. Conservation planning for water quality and soil health also opens the door to subsidies like state and federal funding for cover cropping, which actually increased across Iowa by 22 percent in 2016. "We encourage farmers who own sloped land to farm on the contour, which slows the water down, and to reduce tillage as much as possible. We don't see much moldboard plowing—the kind that makes the earth black—in Iowa anymore," he said, referring to exposed black humus.

Farmers used to control weeds by tilling. Today, they use herbicides. The benefit of herbicides, said Ron Heck, is "we no longer have to spend money on fuel and equipment to disturb the soil. Our crops grow better in undisturbed soil, and the only tillage I do now is to chop up the residue laying on top so that the next crop can be planted and grow up through it." Indeed, Ron showed me tens of thousands of dollars' worth of abandoned heavy-duty tillage machinery piled behind a shed like rusted dinosaur skeletons.

The disadvantage of using herbicides from the soil-microbe perspective is bacterial and fungal diversity decreases, probably from a lack of varied food that is sweated out of different weed roots. And glyphosate, the most commonly used herbicide in Iowa, binds to minerals in soil, effectively competing with the microbes left that would otherwise be delivering those mineral nutrients to plants. Repeated annual glyphosate applications may impair certain kinds of mycorrhizal fungi and nitrogen-fixing microbes. Our system of synthetic inputs produces more food. That's a good thing. The downside is, by undermining microbial diversity, our food is less nutritious and our food supply is less secure.

Once news that I was back in school got around my social circle, I got all kinds of advice. "Never take more than one science at a time," advised my friend Ezra, who hasn't been back to school since the 1970s but said he'd never forget the insanity of taking physics and chemistry the same semester. "My high school biology teacher imparted the most important words of wisdom to me," said Beth, "and I've never forgotten them to this day and still heed them. Never pick anything up by its lid."

I also encountered a lot of longing about going back to school on Facebook.

I was in complaint mode. I posted pictures of my trash can full of futile efforts to answer Professor Rubenstein's inheritance problems and whined about how this was all so off-topic for me. But my friends responded with what they would want to study if they could carve out the time/money/bandwidth: art history with a specialization in illuminated manuscripts, cardiac medicine, meteorology, marine biology, fashion design, archaeology, geology, biochemistry . . . it was pretty dazzling and totally humbling. I had no idea these interests were simmering beneath my friends' cocktail napkins.

Their buoyant certainty is in stark contrast to the insecurity and pressure college students seem to feel trying to decide what they are going to be, according to a blog post on the *Psychology Today* website. It's one thing if you know at a young age, but it's quite another to have to choose a career path. It's a fantasy question that suddenly becomes hardcore reality upon graduation. Students feel the pressure to find the perfect job that "does it all: engages our talents, makes us plenty of money, and reflects well on our family," wrote Brad Waters, a career coach and blogger. But if a revelation doesn't come by sophomore year, when most students pick their major, they can always go for a career assessment.

Career assessments are part of the counseling toolkits some colleges use to help students pick their majors. I took a couple of assessments—including a freebie on the internet that was ultimately about steering you to one of its sponsored for-profit universities—just to see what college kids face. How I answered questions like "Do you think that almost everything can be analyzed?" told me I was an extrovert, was someone who preferred intuition over sensing, was judgmental versus perceptive, was thinking over feeling, and had a tendency to be "larger than life" in describing my projects or proposals—a pretty embarrassing assessment if you ask me. I think I sound like an asshole. Indeed, "this ability may be expressed as salesmanship," my report read, "or stand-up comedy." Disconcertingly, the career for my personality type was politics.

Parents definitely play a role in influencing a child's career choice. Sometimes we do it by holding the education purse strings. When I was a young student, I knew a sophomore who pined away in premed, as per his parents' demands, yet his passion, he confessed, was folk dancing. Or it can be economic. I've met orthopedic surgeons who longed to be in soap operas and accountants who dreamt of starring in *Cats*. Some research suggests career choice may be genetic (as when a child from musician parents shows musical talent) or launched by something the child sees (inspired by the movie *Amadeus*, he decides to become a composer). But whatever path the kids in my biology class take, according to research published in the *Journal of Business and Psychology*, millennials are more likely to make choices based on lifestyle preferences, emphasizing where they live over a particular job, time for personal stuff, and if they can make a difference.

Many millennial values align with the homesteading movement, the DIY lifestyle that promotes greater self-sufficiency, a small energy footprint, and local food. My son's friend, who was raised in Manhattan, is planning on going into organic farming, in order "to help change our overall farming system from one that is resource intensive and environmentally inefficient to one that more closely resembles nature (i.e., high biodiversity) and is therefore more sustainable and also provides a lot of labor opportunities. I also love food and think that being able to provide the majority of food for myself would be cool."

So how is Jasper going to make that a reality, assuming he can afford to buy land with agricultural potential? What methods restore microbial diversity in soil? Farmers have a long history of adding inputs to the soil, like bonemeal as a source of phosphorus, wood ash as a source of potassium, and livestock manure as a source of nitrogen, without realizing why it worked—just that it did. In the nutrient-poor Amazon, for example, pre-Columbian farmers improved the soil by adding charcoal, bone, and manure. (The current incarnation of this practice is called biochar.) The organic movement arose in the early 20th century when some scientists questioned the wisdom of chemical inputs and turned to ancient farming cultures for answers.

Sir Albert Howard was a 20th century English botanist and mycologist whose research convinced him that composting was key to productive farming. Composting is like giving the soil microbiome a fecal transplant, just like the Egyptian shovels of fava-field dirt, and composting cults are legion. The hugelkultur method is a centuries-old farming technique used in Germany and Eastern Europe where plants are grown on mounds of woody debris and compost; vermiculture is creating compost by means of earthworms; Allan Savory's holistic grazing hypothesis seeks to reenact the movements of prehistoric herds and their manure to counter desertification and support grasslands; and the Japanese bokashi method uses anaerobic microbes to ferment waste.

Sir Howard described the principle of the "law of return," that nutrients removed from the soil by crops must be returned to the soil by other means. His books influenced many others, including J.I. Rodale, who, according to organic farming advocate Grace Gershuny, "cranked out a continuous supply of information on the subject since 1942."

Another was Eve Balfour. Balfour was an English earl's daughter with the education and the resources to create, in 1939, the first ecologically designed agricultural research project. Her study observed two adjoining farms composed of temporary pasture and arable land. Each farm housed a herd of dairy cows, a flock of poultry, and a mob of sheep that were fed by crops grown on the farm from homegrown seed, and all crop and animal waste were composted and used on the farm. They were, in essence, identical. However, one farm also received supplemental chemicals (like nitrogen and phosphorus) as well as herbicides, insecticides, and fungicides. The other farm was entirely dependent on its own biological fertility.

In contrast to orthodox agricultural research based on randomized small plots, Balfour's idea was to look at the biological interdependencies of the whole farm. And this is important, because more can be understood about microbial interactions if they are observed from an ecological standpoint. She found out that on both farms, levels of available minerals in soil fluctuated according to the season, with maximum levels coinciding with the time of maximum plant demand. These fluctuations were significantly more marked on the organic farm, where there was more nitrogen and more phosphorus in the soil compared to the farm that had used supplemental nitrogen and phosphorus, suggesting microbial sequestration was at work. She also noted less insect damage on the farm that didn't use pesticides and concluded this was the result of a robust soil microbiome providing nutrients and immune functions on behalf of the plant. Balfour was one of a handful of pioneers with one thing in common, she said. They were "what we should now call Ecologists. . . . They looked at the living world . . . [with an] understanding that all life is one."

But the earliest pioneer in the movement, she noted, was Rudolf Steiner, an Austrian-born philosopher who, in 1924, developed the concept of biodynamic agriculture. Biodynamic farming views the farm as a discrete and potentially self-sustaining organism. Manure produced by animals on the property fertilizes the soil in which the feed and vegetables are grown. The farms use no synthetic inputs and no mechanized irrigation, which slows water runoff. This is not really the same as organic agriculture, which tolerates inputs in the soil like pyrethrum, a toxin from the chrysanthemum plant that kills insects. In the biodynamic system, fertility of the soil is seen as a symptom of the good health of the whole farm, and diseases of plants and animals as ill health of the whole farm. Steiner also called for fertilizer preparations like yarrow blossoms stuffed into a stag bladder, cured and sprayed on plants or buried in the field. It's kooky, but the basic idea is, add microbes.

"Biodynamic preparations are about amping up the life force," said Lance Hanson of Jack Rabbit Hill Winery in Colorado, an organic farmer who switched to biodynamic practices. "The conventional organic is about inputs substitution. But the biodynamic preparations stimulate complexity in the soil"—he's referring to microbial diversity—"and the more complex the soil, the more vital the plant, the more resilient it is. If we have a mite that is overwhelming the vineyard, that's a symptom that tells us we don't have enough things in the environment to control them. The solution is always to build more microorganisms." (The downside of the biodynamic approach is it's based on the premise that there is "a natural tight balance in nature," said the ecologist Timothy Crews, and to a degree this is true, but sometimes there are surges of particular organisms that are "more dynamic than symptomatic.")

The Food and Agriculture Organization of the United Nations estimated in 2015 that 33 percent of global soils were degraded and said this trend must be reversed through sustainable soil-management practices. But how do we get there *and* feed the masses? The Land Institute's mission is to solve the problem of soil degradation by developing strains of perennial grains that allow for seasonal harvesting. This would leave intact the belowground symbiotic relationships between plants and soil microbes that counter erosion, pesticides, and lost yield due to extreme weather. No-till and cover cropping are tweaks, said Dr. Crews. "It's not really addressing the depth of the problem. Deep roots and undisturbed soil. It's the only way."

To that end, the Land Institute is trying to breed new varieties of crops like wheat, sunflower, and sorghum using artificial selection (every year they select the seeds from the plants with the characteristics they desire) and hybridizing perennials and annuals to create plants with the food characteristics of annuals but the ability to grow like a perennial. It's selection on a deadline, and it isn't easy. All our food started as wild plant populations. The grains we eat now have been under cultivation for 10,000 years. In the past 2,000 years, we've only domesticated a small number of wild plants. "We've got a lot of tools our Neanderthal ancestors didn't have," Dr. Crews explained. "But it's still a slow process and slower than our society is used to dealing with. We think it will take about 50 years to take over wheat or rice lands with perennials." (The Institute's Kernza, a perennial wheatgrass that makes pretty good bread and great pancakes, and perennial rice are, according to the institute, already being grown in China.)

In the meantime, farms could be adapted to mitigate the problems of erosion and hypoxia. Farmers could replace synthetic fertilizers with organics and monoculture with diverse crop rotations, plant local breeds, and use cover crop protection. Government could level the playing field by ending subsidies. When it comes to informed choices, like how we are going to farm and how we are going to support our farmers, it seems like it is just as important to understand in what ways we are disconnected from microbes as it is to understand in what ways we are connected.

I knew I was in trouble as we neared the end of the semester and I still couldn't solve a single Punnett square. Professor Rubenstein had supplied us with study sheets, but he pointed out we were also responsible for all of the terminology and material in 20 hours of lectures, 20 hours of recitation, and 18 chapters of textbook consisting of almost 300 pages of incredibly dense stuff. But none of that fazed me compared to probabilities.

In anticipation of the test, I requested a private tutorial with my TA, and after she gamely tried to explain how probability worked, she finally just gave me a few tips that might improve my odds on the test. Ironically, the only probability I managed was calculating how many questions on a test I had to answer correctly in order to pass. In my journal, I wrote about how I felt like I was betraying my learning self. In reality, I didn't give a damn. I just wanted this class to be over. The morning of the test, Professor Rubenstein posted on Twitter, "I love exam day in Intro Bio because there are so many new faces . . . ones who have never been to class since I started teaching weeks ago." This was promising. Maybe the fact that I was never even late might help my grade.

As the tests were handed out, Professor Rubenstein said to answer the math problems on the back pages first, so that we'd have enough time to show all our calculations. I went right to the last three pages of the 12-page test. I calculated, I erased, I recalculated, I erased. I moved to another question, I calculated, I erased. My eraser started to wear down; I took another pencil out of my backpack, which attracted a TA's attention. I waved the pencil to prove my good intentions. I erased.

And then Professor Rubenstein said, "Fifteen minutes!"

Fifteen minutes! WHAT? I flipped to the front of the test, which also dealt with genetics but in sentence form. I whipped through the true-and-false questions—true, false, true, true, false, true—then on to multiple choice—A, C, B, all of the above, none of the above—then to match and fill-in-the-blank. I was almost, almost at the end, and then he announced, "Time's up."

In the subway on the way home, my mood darkened. I couldn't do anything but obsess over what I knew was a rout, a washout, a total bomb. I sat in the train, muttering to myself, "I hate genetics. I hate genetics." But I knew I was being ridiculous. I was hating on a scientific subject with no small implications in my own life because I couldn't face defeat. It was like organized sports. I sneered at them in high school, claiming they promoted competition, inflamed antagonism, and thwarted individualism. But actually, I hated field hockey because I was small and timid and always picked last for teams. Maybe I just lack sporty genes.

The train was full of SantaCon revelers—young men and women in various stages of drunkenness dressed in bedraggled Santa Claus outfits, taking selfies and dropping their phones. Christmas Break had begun.

Shortly after New Year's Day our grades were posted. To my relief, I got the famous Columbia B. But I couldn't help but wonder what my grade would have been had I gone to City College.

CHAPTER

8

The Plant Microbiome

O ver the course of the holiday, my attitude about the Columbia B
evolved. At first, I felt like the beneficiary of a conspiracy, but the
more I thought about it, the more the grade seemed to say, it's okay
with us if you try again. And as the Christmas holiday stretched out, I started
to miss that exquisite feeling of discovery, those little pricks of awe as I glimpsed
the complexity of a neural cell, a drop of water, a teaspoon of soil—those
insights that made the nape of my neck tingle. I couldn't give that up just
because I was unable to divide favorable outcomes by possible outcomes.
Besides, I was rested, and like with pregnancy labor, my memory of the pain
was fading. A new year was approaching and I felt ready to try again.

I put my failures of character behind me and signed up for another round of
Environmental Biology. Semester two would teach us an integrated view of life
within the Earth system, from the minute chemical workings of photosynthesis
to population ecology. Which meant new professors and new styles of teaching.

That's why I decided to hire a tutor.

Years ago, when my kids were still in grade school, I used a local New York
service called Thinking Caps, run by an entrepreneurial gal who once was my
daughter's chemistry tutor and went on to provide both my children with SAT
prep. Most students in New York City get SAT tutoring, whether it is provided
free by their school or hired out. There's one tutoring company based in downtown Manhattan that promises a 430-point bounce from the practice SAT to
the official SAT and costs $1,000 an hour, for a minimum of 21 hours. Parents

who pay these hefty sums prefer an Ivy League graduate to provide test-prep for their kids—and here's where it gets ironic—so their kids can get into an Ivy League college, which prepares them to become tutors.

It had been about 4 years since I had been in contact with Thinking Caps. Luckily, Alexandra remembered me. "I reached out to a couple of our upper level bio instructors and came up with someone I think is fabulous," she wrote.

Jonathan was indeed fabulous. A senior undergrad in neurology, he was tall and thin, a fan of pressed button-down shirts and classical music, and sporting the orange complexion of a vegan. "It's a pleasure to teach someone who doesn't get upset when they realize it's time for tutoring," he said when we met. I told him I needed help with, well, everything, but particularly with chemical systems like photosynthesis. "No problem," he said. "My 9th graders are doing the same material."

It's strange to say, but I found that comforting.

On the first day of second semester I was, as always, early. Our classroom was in Hamilton Hall again, the 1907 building designed by McKim, Mead & White (the architects who also built New York's historic Pennsylvania Station, torn down and replaced by an underground station surmounted by the sleazy Madison Square Garden). Hamilton Hall is copper-roofed and stately, though curiously lacking in a clear bathroom plan. I actually had ongoing problems finding commodes on campus, which I assumed was some kind of middle-aged problem, until I learned that the bathroom patterns in Columbia's academic buildings are weird enough to engender comment in the student news blog. It explained, in its own insidery twenty-something way, how to find the restrooms in Hamilton.

Floors one, three, five
Women, find your place to jive.
Floors one, three, seven
Dudes will find your personal heaven.
(In the vein of woooh!!) Fuck the gender binary!!!

Class was held in a small lecture hall with tiered seating and two big windows that looked out over the snowy hedges of an adjoining formal garden. I took a seat near the front that, upon retrospect, was almost the same seat I took in another classroom the semester before. As the rows filled behind me, the snow on the students' boots slowly melted and chilly puddles crept down between the seats. The majority of students my first semester were freshmen or first-year postbaccalaureates, and I recognized many as they filed in to choose a seat for the rest of the year. They seemed more comfortable with being in school this semester. The earnest young man who had asked if I was a piece of ricotta pie now sat in the back with a row of buddies. There were friendships and alliances all over the classroom.

I was feeling more at ease with being a student, too. I had my tutor in the wings. In the past, I'd rummage through my black bag looking for my black wallet in which I'd placed my ID along with about 20 credit cards while students waited behind me, eventually pulling their phones out of their pockets—the universal 21st century sign of being resolved to wait. Now I kept my student ID handy so I could swiftly swipe into different buildings. I was, finally, a whiz at getting water out of the water dispenser in the cafeteria. I'd spent an inordinate amount of time holding my cup under the faucet, pushing the lever back, to the right, to the left, stepping away to see if someone else might fill a glass as an example, but it remained a mystery for weeks until finally I asked a young woman who wasn't wearing ear buds what the secret was, and she said, "Up. You press the cup up. It took me my entire freshman year to figure it out." After drinking nasty diner coffee and falling behind in household errands, I identified the neighborhood Starbucks and started to take advantage of local vendors: the FedEx office, the shoe repairman, the grocery store. Most of these were on the extortionate side of expensive. The only cheap thing you could get around Columbia was a glass of beer.

The class vibe was rowdier. Happier. Maybe it was because we had turned the corner on winter, though it was still snowing outside. One gal I recognized, a post-bac like me, took off her coat and sat nearby and asked how I did last semester. I told her I got the B, and she gave me a sympathetic look.

I felt a twinge of annoyance. Plenty of students have pretty strong feelings regarding the protocol of asking how you did on a test. I asked Jonathan, my tutor, and he said if people are asking about your grade for productive reasons, like how to master the material better, "I'm cool with it. But for ego reasons? Eh." I sent my college-aged son a text requesting his thoughts about kids that ask how you did on a test, and his response was a virtual epic compared to his usual monosyllabic answers. For the girls, he wrote (and he was referring mainly to high school), it was often out of pretentious self-validation, for the guys, more of a way to joke about how badly they did. "As soon as there is a letter grade, people assume it means the difference between smart and dumb. But I learned early on that isn't true, because total morons who lacked basic logic were acing spelling tests." Some students pretend not to do well for social reasons—they want to fit in*—and others for more shrewd motives. In Chinese culture, it is a strategic advantage to be known as less able than your rival. That's the source of the phrase "crouching tiger, hidden dragon."

Anyway, my fellow student got an A-, but out of either kindness or acknowledgment of the Columbia brand, she added, "But I think I earned more like a C+."

As in the past, the class hushed and inspected our professor carefully as she walked in. Natalie Boelman was slender and tender, the sort of person you'd be comfortable asking to hold your place in a line while you ran to put change in the parking meter. She had young children, and I think that bubble of intense family life influenced the way she communicated. She started her talk with the usual summary of expectations, office hours, and so on, but her graphics seemed lifted from a Scholastic magazine. It's not an easy job explaining bioenergetics, respiration, and photosynthesis, but honestly, it doesn't make it easier just because the images of energy are represented by a sun with a smiley face and coenzymes as yellow taxicabs.

Professor Boelman is an ecosystem ecologist at the Lamont-Doherty Earth Observatory at Columbia University, which is on the west side of the Hudson River, in Palisades, New York. She analyzes how climate change affects the interaction of plants, insects, and migratory songbirds. During one class, she veered into her research to illustrate the importance of coordinated feeding, called trophic synchrony, which is how certain things need to happen at certain times for everyone concerned to be fed.

For example, global warming is causing spring snowmelt on the Arctic

---

*During his presidency, Barack Obama remarked that self-sabotage in minority classrooms, where students who receive good grades are perceived as "acting white," was a matter of national concern.

tundra to occur earlier, which makes plants like berries ripen earlier and causes insects to emerge earlier, too. That's a problem because American songbird migration to the tundra, where they hatch their eggs, evolved to happen when the berries and insects are available. But with global warming, the birds may be out of sync with the emergence of their food source. In a blog post Professor Boelman wrote about trophic synchrony for the *New York Times*, she quoted her colleague Eric Post. Imagine that you go to the cafeteria for lunch every day from noon to 1 p.m. One day the cafeteria decides to start opening at 11 a.m. and closing at noon, without letting you know. When you show up at noon, you might be lucky enough to score some leftovers, but it's also possible that you'll completely miss out on lunch. (Add to that the fact that you flew thousands of miles to get to the cafeteria.)

But mostly Professor Boelman's classes reflected her day job, teaching us the essential systems of biology and, in particular, the biology of plants. And there is no deep or complete understanding of the biology of plants without understanding their microbiology. Actually, you can't understand plants without understanding the microbiology of soil, too, because it turns out that plants and soil are connected by microbes.

"A rose is partly a rose, but mostly it is lots of other things," wrote the mycologist Nicholas Money in *The Amoeba in the Room*. "Its roots, leaves, stem, thorns, and flowers are caked with microorganisms, its inner anatomy is riddled with other life forms, and the whole magnificent botanical enterprise of the rose is sustained by a medley of soil bacteria and fungi."

It was when I was studying plants that I experienced one of the most profound realizations I had about biology. Every higher organism has a microbiome, multiple microbiomes, in fact. The different microbiomes of a plant or a person have evolved to fill a niche—as if the host was a landscape—and the role the microbes play in that niche can be so important that without them the host fails to thrive. The microbes in microbiomes negotiate the acquisition of their host's nutrition, its immunity, and its ability to adapt. They function at the interface between the host and the host's environment.

They are of us but apart from us. They help us navigate the world, and we are their world.

Imagine a plant and its microbes not as a city with its own citizens, but rather a country with multiple cities, each with their own citizens. Plants have a microbiome at the plant-soil interface (rhizosphere) and another at the air-plant interface (phyllosphere). But there are also microbiomes of the antho-sphere (the flower),* the spermosphere (seeds), and the carposphere (fruit). This is the same for humans. We have a gut microbiome and a skin microbiome—those are the big ones—but also a nose microbiome and an ear microbiome and a bellybutton microbiome, to name a few.

Just to get the syntax right, when scientists refer to a plant (or human) microbiome, they are referring to all the microbial communities that live in and on the host. But when scientists refer to a specific microbiome, they'll identify it by location, like the root microbiome of a corn plant or the gut microbiome of an accountant.

All microbiomes of plants below and above ground include bacteria and fungi, as well as other critters like archaea, viruses, and insects. Some live on the plant (ectophytes, from the Greek *ecto* for "outside" and *phyte* for "plant"), others live in the tissues, between and even inside the cells of the plant (endo-phytes, from the Greek *endo* for "inside"), and both can live concurrently in a particular part of the plant. These microbes are party to ancient relationships that help plants deal with their main problem—they can't move. Because plants can't run away, they have to find other ways to deal with predators, and since they can't relocate, they have to adjust to local environmental conditions, and because they can't forage beyond their immediate vicinity, they need help get-ting enough of the right nutrients. "The plant, in short, is little more than a photosynthetic factory with problems of distribution and supply like any man-ufacturer," wrote the paleontologist Richard Fortey in *Life*. "I wonder if it is a coincidence that complex and large factories are referred to as 'plants.'"

A plant's microbiome is a toolkit for survival. When a plant needs the help of a specific type of microbe, it recruits the fellows it needs from the soil by lur-ing them with a particular secretion they can't resist. That's why microbial

---

*The anthosphere is a natural hub for bees, which gather more than pollen. They pick up bacteria needed in the bee microbiome.

diversity in soil is so important. Plants need a variety of local microbes to recruit from. A plant is like a company that counts on a pool of temps with variable skills that they can tap when their business needs change. For example, in a study of tobacco crops, sudden-wilt disease emerged but was suppressed by a consortium of bacteria that weren't a regular part of the plants' microbiome. Like a platoon of mercenaries, they were hired to fight the disease.

Plant microbiomes are regionally sourced. They come from the neighborhood. Plants inherit some microbes from their parent when they are seeds; others are delivered by insects, wind, or rain; and still others are drafted by the plant from the soil, as quickly as within 24 hours of sowing. The microbes that colonize a seedling aren't killed by the plant's immune system, because the ability to recognize one's symbionts is innate. They're part of who the plant is.

In fact, the regional sourcing of beneficial microbes is consistent in nature. A baby picks up his initial microbiome from Mom on the way through the birth canal. More populations of microbes are acquired when he starts eating food and even more from his sisters and his dog and his hometown, farm, and forest. Our immune system doesn't attack them because it recognizes they are part of us.

Since microbes are recruited from the environment, what kind of microbes live in a particular species of plant—or for that matter, in a particular people— is dictated by their environment, including the geography, the altitude, the water source, the weather, and the soil microbes that have evolved to live there, too. However, the plant determines the makeup of its microbiome. The plant selects how many of whom can move in, and as the plant grows, it acquires more and different microbes, as needed. And once again, the same is true of other complex organisms, like my friend Jimpa and me.

I once visited a dumpling house affiliated with the Lamaling Monastery in Tibet. From a distance, it's a wonderful place, with huge wooden beams and thick wooden tables and wide wooden bowls filled with steaming momo dumplings and big thermoses of hot tea. It was packed with people, children scampering about, wrinkled old women wearing thick coral beads and smoking pipes. Or maybe they were men. Upon closer inspection, it was incredibly filthy. I mean, totally disgusting. When people finished eating, they brushed their garbage on the floor, and then a couple of dogs would suddenly appear from wherever they were lurking and start snarling and fighting over the tough ends of the momos and the spittle-filled tissues and the spilled sepen, the local hot sauce. I ate one momo, a gray mystery meatball housed in a tough doughy skin and boiled, and pretty quickly my stomach started to complain, but our interpreter Jimpa happily tossed down eight or so of these grimy dumplings. On the drive out of the valley, my stomach continued to whine. Jimpa, in contrast, slept like a well-fed baby.

The population composition of the microbial community in me is determined by which microbes I am exposed to and which ones I feed. The reason I had a bad reaction to the momos (and why I get the runs when I drink the water in Mexico) is the bacterial colony in my gut reflects the foods I regularly feed it, and the introduction of a new species of bacteria on a food, or a food that my resident microbes are unfamiliar with, can throw my digestion off. If I kept eating momos in Tibet, and survived, I would eventually change the population of my gut bacteria to one that can handle momos. In other words, I would be selecting my microbiome, just as plants do.

Some microbes in plants are transient. They're temporary visitors who may or may not provide some benefit to the plant. Others are permanent residents. They've found a niche in the plant, and they pay their rent with some service that the plant benefits from. These microbes are part of the normal flora of the flora. Some microbes in us are transient, too. When I shake your hand, some of my skin bacteria get on you. On me, the bacteria are permanent residents that play an important role on my skin. On you, they are transient.

On and in any given plant, you'll find an array of microbial lifestyles. There are microbes that are not causing any trouble but not really doing anything helpful either, just hitching a ride on the plant. You'll find pathogenic microbes that can cause disease if there are enough of them and mutualists that mainly help the plant get nutrition or defend it in exchange for the plant's scrumptious secretions. Same goes with us. We house freeloaders, and microbes that could do us harm if their population numbers increased, like the biofilms that cause dental plaque, and mutualists who pay their way with the currency of nutrition or defense.

Microbes that live in plants make chemicals and nutrients that plants can't make but need, like hormones that regulate plant growth or chemicals that aid in reproduction, like furaneol, a tasty chemical in strawberries that attracts the birds who spread the seeds, and red anthocyanin, the pigment that makes them colorful. Certain bacteria produce chemicals that can act like a switch to get the plant to start doing something healthy, like using the nitrogen other bacteria have fixed. Plants also eat bacteria. My biology book explained most plants were autotrophs, that they made their own food. But maybe the carnivorous Venus flytraps, which get their nitrogen from bugs (they grow in places without a lot of nitrogen in the soil), are not the carnival oddities I grew up thinking. Maybe this is the norm, just on a really gaudy scale, because researchers in Australia have discovered that plants can degrade a bacterium while it's inside the cell and make a meal of its vitamins, nitrogen, and proteins.

A mix of bacteria, archaea, protists, and fungi plays these different roles in plants, but fungi play a particularly significant part. In general, soil fungi provide nutrition directly to the roots of plants, but *inside* the tissues of the plants, they provide protection. All terrestrial plants probably have endophytic fungi wriggling between their cells, many passed from one plant generation to the next aboard seeds. These fungi have diversified into a million species that have evolved to coexist with different plants (though most are either Zygomycota, Basidiomycota, or Ascomycota), dining on the products of photosynthesis. In return, they're like Prozac for plants.

Here's how. When you smoke a cigarette, it can cause something called oxidative stress. An increase in free radicals (aka unpaired free electrons) attack structures like DNA if there are more of them present than your cell can handle. That can lead to cancer. Your body mitigates the effects of cigarette smoking by producing antioxidative stress chemicals to counter the cell damage, as do certain plant foods like kale. Similarly, when a plant gets stressed from drought—as if the plant smoked a cigarette—it suffers from oxidative stress, which can hurt its cells. But unlike you and me, plants don't produce those helpful antioxidative chemicals to counter the effects of drought. It's the endophytic fungi living inside the plants that do.

These impossibly thin fungal threads emit an arsenal of compounds that calm the oxidative stress of plants. They also participate in the chemistry that makes plants use water efficiently, which helps not just plants with a drought problem but also plants stressed out by heat or salt exposure.

"These are ancient symbioses," said Russell Rodriguez when I met with him at his lab in Seattle. (His organization, Adaptive Symbiotic Technologies, develops solutions to agricultural problems using fungal symbionts.) "Every endophyte we've studied confers drought tolerance." Endophytes that help plants deal with drought and conserve water are, along with mycorrhizal fungi, prevalent and ancient plant symbionts. They helped plants manage life out of the sea in the first place. About 5 percent or so of plants lack persistent fungal symbionts, but they aren't even living on land anymore. Without their fungal partners, those plants returned to being aquatic again.

Endophytic fungi also defend plants from herbivores by means of a virtual armory of deterrent toxins, from bitter flavors to more lethal compounds like alkaloids, which are basically poisonous.* For example, an endophytic fungus produces the active ingredient in locoweed, swainsonine, which is toxic to grazing animals. The endophytic fungus *Epichloë coenophiala* lives in fescue grass. It produces loline (and other alkaloids) that are toxic to herbivores, which is why fescue grass is a much better choice for golf courses than pony pastures. There's actually more to this story (there always is). A study from the Microbial Ecology Lab at Southern Connecticut State University found that the loline is produced in quantities relative to the threat. If the grass has a bite taken out of it, the endophytic fungus, which is also bitten, responds by increasing the quantity of toxin. The plant releases excess loline into the soil around the roots, where it attracts a community of nitrogen- and phosphate-fixing bacteria that promote the plant's regrowth.

The mere presence of endophytic fungi provides an immune service. They seem to irritate the plant, and in the process, they activate the plant's own immune system, but not against the endophyte. The plant tolerates the endophyte and, with its primed immune system, is better able to ward off real threats. That's similar to the rationale behind the hygiene hypothesis, which

---

*Though I should note that plants produce alkaloids as do endophytic bacteria. Caffeine, tobacco, and morphine are plant-made alkaloids. They also produce defensive phytochemicals. Capsaicin, the stuff that makes a pepper hot, is a fungicide.

says the rise in human immune diseases like asthma is a result of underexposure to the bacteria that prime the human immune system. Plants that are exposed to fungicides that kill their endophytes could hypothetically develop an immune disorder.

That got me wondering if certain farming practices, like using fungicides, affect more than the microorganisms in soil, but the microbiome of the plant, too, even the nutritional value of a crop? It turns out the nutritional content of the plants we consume is related to the farming practices used to grow them, and in particular, how those practices affect the microbes that live in and on crops. When I went to Iowa to poke around in the soil at Ron Heck's farm, I thought I'd see if the way corn was grown affected the corn itself. Plus, the week I visited the Heck farm, the Iowa State Fair was going on.

First held in 1854, and at the fairgrounds in Des Moines since 1886, the Iowa State Fair is huge—445 acres including a campground—and attracts a million visitors over the course of 9 days. The largest agricultural fair in the world, it has hosted acts like Sonny and Cher, Johnny Cash, and KISS, among others. The fair is a necessary stop for presidents and presidential candidates. Just like in the microbial world, the fair houses its share of temporary visitors who may or may not provide some benefit to the overall enterprise.

When I visited, there were tens of thousands of exhibits, from hogs to fabric napping, and vendors selling everything from hair extensions to hot tubs. There was a heritage village with a restored one-room school house and post office, a swine barn filled with fat pink pigs, a horse barn with massive Clydesdales, and a tremendous, hot cattle barn that housed, among others, a set of six identical cows that hyped the farm's genetics and the 2,900-pound Desperado, a bull that looked like it might have ranged Iowa during the Ice Age. I stopped to see the refrigerated display of a life-size cow made of butter, an Iowa State Fair perennial, as well as the deck of the Starship Enterprise, with Captain Kirk about the size of a husky seventh grader, the whole cast portrayed in pale yellow butterfat.

This is patriot country, where 100,000 people will stop drinking beer to listen—in stillness and reverent silence, hands over their hearts and facing the nearest loudspeaker—while the Star-Spangled Banner plays over the PA. In the pavilion, heritage families were awarded plaques celebrating 100 years of farming, including Ron Heck and his family of four, but some families were gigantic, with as many as 18 members, spanning the mobility arc from strollers to walkers. I visited the centennial event in the morning, stopping first at a tent to eat a blueberry pancake that was bigger than the plate it was served on. I ended up chatting with an unsmiling couple that was also receiving a centennial farm award. "What county are you from?" they asked. "Manhattan," I answered, but they didn't think it was funny.

If there was a theme, it was corn. The most prevalent crop in the horticulture exhibits was corn: hybrid commercial corn, white, yellow, longest ear, open-pollinated corn, ornamental corn, popcorn, and corn sculpture. All the animals (except maybe the rabbits) were fat and glossy on a diet of corn. The bulk of the food vendors were selling corn-based foods, like corn on the cob, corn chips, and five types of corn dog including gluten-free. If Iowa were a country, it would be the third largest corn producer in the world. Most of Iowa corn is field corn used for ethanol production and livestock feed, but it is also used to make starches, sweeteners, and thousands of everyday products like lipstick, tires, and plastic water bottles. Only 1 percent of Iowa corn is sweet corn, the corn used for canning, freezing, and eating fresh. It is, however, grown much the same way as field corn.

Corn is farmed in association with a variety of chemicals: fungicides, pesticides, herbicides, and synthetic fertilizers. It's genetically modified to manage these chemicals and have certain physiological characteristics like size and sugar content. All of these factors affect the microbiomes of both the soil and the corn plant.

The main causes of corn plant diseases are fungi, though there are bacterial pathogens, as well. Hybrid corns are modified with disease-resistance genes, but what really does the job is spraying fungicide on Iowa fields. Broad-spectrum fungicides are like broad-spectrum antibiotics. They kill the good guys along with the bad. Fungicides that are effective at killing pathogenic fungi can kill both the fungi that ward off insect pests and the fungi that help the plant handle stress.

Additionally, when farmers use fungicides, especially as insurance against an infestation, it leads to fungicide resistance, where the fungus they want to get rid of adapts to the fungicide. In some cases, this can happen in as little as 2 years. Indeed, by using a fungicide excessively, we are essentially selecting for fungicide immunity and creating superfungi, just like our overuse of antibiotics is creating antibiotic-resistant superbugs.

Iowa farmers also spray their fields annually with 30 million pounds of herbicide, mostly glyphosate produced by Monsanto, a multinational agrochemical company that started in food additives and then moved on to PCBs, DDT, Agent Orange, and other products we have since come to regret. Originally invented to clean dishwashers and other appliances, glyphosate is a degradable herbicide sold under the name Roundup.* Roundup kills most plants it encounters, but not those that have been genetically modified to resist the herbicide. (A bacterial gene that codes for an enzyme that is immune to glyphosate is inserted into the plant's genome.) Most nonorganic corn in Iowa is grown from "Roundup Ready" seed that is glyphosate-resistant. Monsanto produces both the herbicide and the herbicide-resistant seed.

Glyphosate kills plants by inhibiting an enzyme they need to produce certain kinds of amino acids. Since we don't make those amino acids but get them from our diet, the chemical doesn't affect us. Nonetheless, it is showing up in our gut, where it may affect unexpected targets like our gut microbiome, and we don't know what impact that will have on us in the long run.

There are always repercussions when an ecosystem is tampered with. The symbiotic microbes on weeds will do everything they can to survive the glyphosate, and if you repeatedly expose a plant to the same herbicide, those microbes will help that weed find a way to survive. "Microbes are responsible in part for the hardiness of these weeds," said James White, a mycologist and plant pathologist at Rutgers University. How are we dealing with this? New glyphosate herbicides are being developed to "build on" Roundup.

Currently, glyphosate modifications are present in 88 percent of corn grown in the USA. It's the most current incarnation of a 9,000-year-old project.

---

*Though how degradable is open to question. A paper published in the journal *Environmental Health* suggests glyphosate takes longer to degrade than Monsanto says. And if claims it is carcinogenic turn out to be true, that's a problem.

Corn (or maize; they're the same species) has been genetically modified by artificial selection since the Neolithic Era. Today's European hybrid was originally
teosinte, a Central American grass with an inch-long seed tassel. In an incredible bit of archeology in 1948, a couple of Harvard grad students exploring Bat
Cave in New Mexico, a site near ancient agricultural fields, found a history of
corn cultivation. The deeper they dug, the smaller and more primitive the cobs
became, until at the bottom they found pod corn, the earliest form of corn discovered, in which each individual kernel was encased in its own husk.

In 1896, agricultural scientists at the University of Illinois began selecting
for the oil content in corn kernels. The experiment continued for decades, and
after 90 generations, the average oil content had increased by 20 percent. Other
breeding programs have created corn plants that are 10 feet tall with up to
20 leaves, displaying nearly a square yard of leaf area per plant to light. It's tur-
bocharged, rapidly converting kinetic energy ("That's sunlight," said Professor
Boelman in her photosynthesis lecture) into chemical energy ("That's sugar,"
she added). You can even hear it growing, a crackling, staticky sound caused by
the rapid growth of the stem between the leaves.

We enjoy corn that is genetically engineered in both the lab and the field,
and one technique is not necessarily better than the other. It may seem ghoulish
to insert bacterial genes into the corn genome, but bacterial DNA is naturally
present in most organisms, and no obviously negative health effects due to the
technique have been documented so far. Indeed, the method of bringing about
change in organisms might be less important than the long-term effects of that
change.

One of those unplanned changes is nutritional content. While I was at the
Iowa State Fair, I ate an ear of broiled corn that was achingly sweet. That's the
work of a geneticist named John Laughnan who in 1959 discovered a corn
mutation that in one generation "eclipsed old-fashioned sweet corn in the marketplace," reported the *New York Times*. Likewise, in the 1800s corn hybrids
that decreased the kernel's yellowness were favored. I remember my grandfa-
ther, who was from Memphis, Tennessee, saying yellow corn was for mules. But
when the yellow color was bred out of corn, so too was the beta-carotene, which
is metabolized into vitamin A in our bodies. Today, sweet white corn is rich in

sugar and starch but low in nutrients; teosinte has 10 times the protein of contemporary corn.

More contemporary genetic modifications like "Roundup Ready" corn supports conventional (a nicer way of saying industrial) farming practices that undermine microbial diversity in and on the plant and in the soil. When you decrease soil microbial diversity with synthetic fertilizers, fungicides, and herbicides, it undermines the nutritional load in the plant. A comparison of organic versus conventional farming methods shows that conventional agriculture alters microbial communities that move micronutrients like zinc, copper, magnesium, and iron from soil to plant. Nutrients taken up by the body of the plant are not returned to the soil, and as the complexity of the soil's nutritional load and the microbes that can deliver those nutrients decline, so too do the nutrients in plants upon which we all depend. Overall, there has been a 5 to 10 percent decrease in minerals in American-grown fruits and vegetables since World War II. As the philosopher farmer Wendell Berry wrote in his essay "The Pleasures of Eating," "Eating is an agricultural act."

Professor Boelman's slides were excruciatingly detailed. Here's an example. A slide titled Carboxylation vs. Oxidation of RuBP: which one will RuBisCO favor? included cartoons of the cell walls of plants under differing temperatures. It also had parentheticals with definitions of carbonization (carbon fixation), oxidation (photorespiration), and RuBP (ribulose bisphosphate, a 5-carbon organic compound that attaches to CO_2, a step in the process of photosynthesis photorespiration). And there was more. On that one slide, we were informed carboxylation is favored by plants at temperatures of 25°C and oxidation is favored at less than 25°C (sorry, no Fahrenheit in bio class). Plus, there was a line from the textbook copyright that was so small I had to read it in order to know I didn't have to. There were dozens of slides like this, which Professor Boelman explained to us by reading the text off them.

Sterling Education Services warns that the way to lull your audience to sleep is to read from text and not make eye contact and use jargon and

acronyms. I certainly felt the wave of sleep-desire come over me as she described
the chain of chemical reactions in photosynthesis that creates glucose and the
chain of chemical reactions in respiration that harvests energy from that glucose. As a result, I had to spend a lot of time with Jonathan practicing these
chains backward and forward, like an endless Hanon piano exercise.

Jonathan was incredibly precise in his explanations of things like the oxidation of glucose, but his diagrams were so erratically drawn that when I looked
over them after he left, I couldn't interpret his scribbles at all. That's how I knew
he was destined to be a doctor. I eventually had to take the pencil from him and
draw the pathways myself: glycolysis, pyruvate oxidation, Krebs cycle, electron
transport chain. When he described concepts like how foreign DNA is cloned
and then introduced into plant cells to genetically modify them, he'd make a
fist, thumb up, and jab at invisible cells, and for a long time I thought the ges-
ture was a rather insensitive reference to giving a shot—like in the movies
where someone is stabbed with a tranquilizer while being held down by a pile
of police officers—but then another scientist pointed out that's how you manage a lab tool, the pipette. Jonathan described complicated processes like how
eukaryotic genes can be expressed in bacteria cells in the first-person plural.
"*We* genetically engineer bacteria to manufacture human insulin," he'd say, or
"*We* figured out how to do this in 1982. It's brilliant," and his eyes would glitter.
That's when I got Jonathan. He wanted in on it—in on the breakthroughs, the
discoveries, the access to truth. And in my own way, so did I.

Despite the challenges of Professor Boelman's class, I was getting the picture that not just plants and microbes were connected but also plants, atmosphere, and microbes were connected. For example, a rise in temperature can
set off a cascade of chemical reactions in a plant, and as a result, the plant
converts less CO_2 into sugar. Since microbes depend on a plant's sugar, atmospheric warming leads to reductions in microbial biomass in the soil.

The evidence is in. Microbial consortia can be negatively affected by
changes to temperature, soil, and plants. But rather than address temperature
change and the effects of agricultural inputs on soil and plants, the trend
today is to reengineer the plant-microbe relationship to perform better under
our current farming scenarios. We took away certain assets from the plant,

and now we are in the unenviable position of figuring out how to put them back in.

Some companies are looking to breed plants to attract the microbes they need, others are trying to figure out ways to add microbes to plants that need them. Companies in the business of pesticides and herbicides, like Monsanto, Bayer, and Syngenta, are researching potential microbial inoculants that might beef up agricultural outputs, a kind of probiotic for plants that boosts growth, increases resistance to drought, disease, and pests, and reduces reliance on fertilizers, pesticides, and herbicides. RootShield, for example, is a fungal spore spray that gardeners and farmers can use on their disease-free seedlings. Once the fungus is established, it's supposed to compete in the soil with pathogenic fungi.

Some market analysts suggest microbial inoculants will swell into a $4.5 billion business by 2019. The packaging is already in place for "biofertilizers" like nitrogen-fixing bacteria, "bioprotectors" like those microorganisms that attack pathogens of plants, and "soil stabilizers" like mycorrhizal fungi that glue soil particles together and sequester water and carbon. Should they work, they could increase yields and decrease the need for fertilizers and irrigation.

Startup companies like Indigo in Boston have embedded microbes into seed coatings in order to "recolonize the plant with its beneficial endophytes." Monsanto's QuickRoots is a microbial seed inoculant for improving corn's ability to uptake nitrogen, phosphorus, and potassium. BioEnsure, a product from Russell Rodriguez's Adaptive Symbiotic Technologies, is a seed inoculant of endophytic fungi that provides protection from extreme ecological conditions like drought. The fungus is so promiscuous it can be used on corn, rice, soybeans, wheat, tomatoes, cotton, and other crops. When it comes to agricultural success, "microorganisms are the key," said Dr. Rodriguez.

Coaxing resident microbes may actually work better than introducing engineered species. Stress tolerance can be conferred to agricultural plants simply by colonizing plants with the appropriate endophyte, though the fungus must be gathered from a plant under stress, suggesting the endophyte develops the stress tolerance in situ but doesn't lose the ability to confer that tolerance to another plant. "Nature has done all the work," said Regina Redman, vice president of Adaptive Symbiotic Technologies. "We just go to the stressed habitats, isolate the endophytes, and bring them home." And these endophytes are accommodating. The species that allows panic grass to grow in soil temperatures up to 149°F also allows tomatoes to grow in similarly hot conditions. The endophyte that allows dune grass to tolerate occasional salt inundation also confers salt tolerance to rice. Because the relationship between fungal endophyte and plant predates the diversification of plant species, the tolerance that originated in one kind of contemporary plant can be conferred to another. To the fungus who knew their common ancestors, panic grass and tomatoes are the same thing.

But the gold ring of plant biotechnology is nitrogen fixation. Nitrogen is the Iowa corn farmer's biggest expense, and it's a large national expense, too, if you consider the energy costs of making the stuff and the economic costs of nitrogen runoff suffocating the Gulf. The challenge is how to encourage corn to fix nitrogen on its own. "Transferring this symbiosis to non-fixing crops is a huge techno dream," said Toby Kiers, a mutualism researcher at Vrije Universiteit Amsterdam, to *Pacific Standard* magazine. "People have been trying to do it for decades—and failing." But it turns out corn may have housed nitrogen-fixers in its evolutionary past.

One hundred million years ago a gene evolved in some plants that allowed a partnership with a nitrogen-fixing bacterium. There are scientists looking for this gene in nitrogen-fixing plants in hopes they can introduce it into non-nitrogen-fixing cereal crops. But centuries of corn cultivation have altered the plant's relationship with some of its historical symbiotic microbes. For example, teosinte kernels have a husk on them. It turns out the husk is the nursery for the microbiome of the plant. James White's lab at Rutgers University has learned that microbes are typically carried on the surfaces of the seeds of plants. "But with modern corn we have selected against those outer tissues. Many microbes still live on modern corn kernels even though the external tissue is removed, but it may be those communities are less able to do things that could benefit the plant, like fix nitrogen," said Dr. White.

The scientists at the Land Institute have positively identified endophytic nitrogen-fixing bacteria on landrace corn from Hopi land in Arizona—though they don't know the rates at which fixation occurs. It may be that when all the nutrients needed by a crop are supplied by synthetic means, the nitrogen-fixers are simply selected against and eventually disappear.

The seed is the future of the plant. There is an incentive for the species to endow that seed with all the microbes needed to ensure the seedling's success, which includes inoculating the surrounding soil with beneficial microbes. Microbes are nutrient traders, and traders travel. When a seed falls to the ground and germinates, bacteria associated with the seeds will colonize the neighborhood and actually patrol the soil to clear out its seedling's competitors and pathogens. One of Dr. White's projects looks at the microbial interactions of common reeds. When a reed seed germinates, some of its seed-borne microbiome colonizes the seedling, entering the tissues, but others venture into the soil and kill all the dandelions.

Because of the interplay between above- and belowground microbes, the removal of the corn seed husk ultimately affects the microbial composition of the soil, too. And this isn't particular to just corn. The microbiome of cottonseeds is nurtured in the seed fibers, but the fibers are removed with concentrated acids so the seeds don't get stuck in automated planters. In contrast, the microbiome of wheat hides in the seed's crevices. Unlike corn kernel husks or cottonseed fibers, the microbes in wheat seeds are difficult to remove. That might be part of the reason why wheat is such a successful crop. It tenaciously holds onto its microbes to the benefit of both its above- and belowground microbiomes.

The microbiome of a plant affects taste, as well. All of the European wines are made from one species of grape, *Vitis vinifera*. Yet how do they achieve such variety in flavors? Through breeding of subspecies, of course, and environmental factors that affect the plant—sun and temperature and minerals in the soil—but the unique flavors of different wines may also be affected by the microbes in the plant and in the soil. Infection of the grape by different species of fungi has been found to enhance or degrade the flavor of the wine, according to wine aficionados.

In my mushroom club we have done side-by-side tastings of morels that were collected at forest-fire sites, where they can grow prolifically, and morels collected under dying (but not dead) elm. Most of the tasters preferred the morels picked under elm. Mycologist Tom Volk pointed out that fire morels may be less tasty because the high temperatures of a forest fire might have damaged the microbial population in the soil and subsequently damaged any flavors their presence might have lent. Likewise, if you compare a wild maitake to a commercial maitake (*Grifola frondosa*), the wild sample tastes more mushroomy and intense. That may be because the wild fungus is accessing a buffet of nutrients in association with bacteria. The cultivated maitake are on a more limited diet of sterilized hardwood sawdust. Indeed, it could be that branded foods like the sweet Vidalia onion, grown in Vidalia, Georgia, are ultimately branding a microbiome.

And taste is connected to nutrition. A British study suggested that organically raised plants produce higher antioxidant levels (the fungal endophyte's job), which affect flavor, aroma, and mouthfeel, as well as benefit us when our cells are under stress. The microbiome of a crop plant, influenced by farming methods or breeding programs or both, affects everything from taste and nutrition to environment.

As we rounded the corner of Professor Boelman's section, she reminded us that the well-being of any organism is subject to what its ecosystem can provide. Deprive a carrot of a complex ecosystem, and you deprive yourself. Maybe industry will someday engineer a domestic plant microbiome that can replicate the benefits of a wild microbiome. But the challenge is great. There are a lot of interactions to understand. Diverse microbes migrate from the seedling into the soil, colonize the soil, colonize other plants, even kill competitive plants, immigrate into the plant, access nutrients, help the plant use those nutrients, suppress pathogens, produce toxic chemicals to protect the plant, provide stress tolerance, and sometimes get eaten by the plant. Microbes necessary to the existence of the plant even come and go; some enter the plant's cells at the soft tips of the plant roots and push out through the root hairs, in effect, blurring the border between plant and soil.

Indeed, microbes compose an edge that is not an edge but rather a bridge between plant and environment. The closer the microbes are to a plant, the greater their numbers, but they live on a spectrum between plants, as well. You could picture microbes like an aura that erases the hard borders of the living plant, and two plants living in proximity might share homogenized microbiomes, like a living Venn diagram. Plants are not only connected to soil by microbes; microbes also connect them to each other. That's not an easy thing to imagine, much less reproduce.

I asked Dr. White if the same might be true of us, if our physical borders might also be a blur of microbes that connect us to each other and our environment. Do we share an open border connected by microbes with other people?

"Well sure," he said. "That's one definition of a family."

CHAPTER

9

Your Microbiome
Is a Park

I 'd like to say that with the help of Jonathan I started to read newspapers
in the morning again, washed my hair, and had lunch with a friend without feeling guilty. But that's not the case. Because along with the usual
heavy reading load came lab.

It's one thing to learn biology. It's a totally different thing to do lab work.
Learning biology is an absorptive process. Lab work is about production, and in
the 21st century, that means computer-generated graphs and tables and math.

"Ugh, lab," said Jonathan. "It's in a stinking room and takes hours and you
have to write these tedious reports and it's all only worth a single crummy
credit."

Our first lab session of the spring semester was held in a big classroom
where we were introduced to the teacher's assistants. There was a young woman
with a lovely smile and a squeaky high voice and a young man who also had a
very high voice and a tendency toward upward inflection. I remember first
hearing the uptalk style in 1982, with Frank Zappa's song "Valley Girl." "There's
like the Galleria? And like all these like really great shoe stores?" On the one
hand, ending every sentence with a question mark seems to convey insecurity
or a need for confirmation—as James Gorman pointed out in the *New York
Times*, imagine a cop saying, "You're under arrest?" Or a surgeon saying, "So,

first I'll open up your chest?"—but it's more likely the speech pattern is the result of a kind of contagion. If you are around it a lot, you pick it up.*

The third TA was Kyle. He had a strong, clean voice and clear blue eyes, a red-head with lots of freckles, the kind of person you'd automatically hand the map to if you went on a road trip. The three TAs separated into different corners of the room, and we were told to assemble near the TA we preferred. I immediately stepped into Kyle's corner. As did another gal I knew, the one who had been curious about my grade. I asked her why she had joined Kyle's group. "That's easy," she said. "The other two's voices would have driven me nuts."

The first lab, held directly after we'd done our sign-up, was actually a mock lab, to show us what we were in for. We were shown a film about wolves in Yellowstone Park, a famous example in biology classes about how every link in the food chain is necessary for a well-functioning ecosystem. What happened was ranchers demanded the Yellowstone wolves be exterminated because they were preying on their sheep. But with the wolves gone, the elk populations increased. The elk hung around in large groups on the banks of the rivers and overbrowsed the willows. The beaver disappeared because they need willows to survive winter. Without the trees, songbirds declined, and without the beaver dams, so did the fish and ducks. When the wolves were reintroduced in 1995, the elk had to stay on the move. They broke into smaller groups, and their grazing was less concentrated. The willow stands increased in biomass, and the birds and beaver came back, as did the ducks and fish. The riverbanks recovered from overgrazing and erosion. With less erosion, there were more riffles and more pools, in short, more niches for other creatures to find a home in the water. The moral of the story is, every link in the food chain matters, and its effects go up to a geographic scale. Then the TAs showed us a bunch of slides about the cascading effects in coastal waters when sharks are taken out of the equation.

So, I was getting everything, and then they explained how a lab report is written: a catchy title; then an abstract, which synthesizes the most salient points; a hypothesis; the methods used in the experiment to test the hypothesis;

*Same goes for particular words. In my lifetime, I've seen the filler word *basically* rise to the level of *like*.

the results, which include figures and tables; the discussion, where you refer to other scientists' work and try to fit your results into a bigger picture; acknowledgments; references; and appendix. Being a writer, I am used to shaping language to suit my purposes, but I'd never written a lab report before. I started it like I was writing for the *New York Post*—Sharks Are Green!—but soon became bogged down writing tortured sentences like "The presence of predatory species in three coastal ecosystems increased the rates of carbon sequestration by predating those species that destroy carbon stocks." I felt like I was screwing a light bulb into a fixture with worn-out threads. I just wasn't getting any traction. How do you describe multiple changes in an ecosystem that are happening at once? "Take out the shark and you don't just break the food chain," I wrote, "you disrupt the system of systems."

Can I even say that? One student, the second to last to leave, looked at me and said, "I'm pissed. This is so stressful and pointless," before she hit send and emailed her text to the TAs, and I thought now *that's* the Columbia attitude! Your education is a product you are buying, and you can be disgruntled if that product is lacking. But me, I am a mouse. I tried to hurry up so the waiting TAs could get to dinner.

I wasn't graded, but I was informed I had failed to include the methods and acknowledgments, and I realized two things: Writing lab reports is not about the language, or interpretation, or persuasion. It's about reporting data and determining how your data fits into the greater understanding of life. That's why references are so important. They help locate the work in the geography of natural revelations. Each scientist's work contributes words to the sentences we write about life.

The other lesson I learned was ecosystems function as a connected community, and pretty much the same rules of community ecology exist whether you are talking about Yellowstone Park or the microbiome of your cousin.

When the wolves of Yellowstone were pulled from the ecosystem, there were cascading effects that undermined the park as a whole. It's the same on a microbial level. When population numbers of microbes are reduced or increased in an ecosystem, the ecosystem itself can get out of whack, sometimes just a little, but oftentimes a lot. When population changes have a debilitating effect, it's called dysbiosis, from the Greek *dys* for "bad" and *biosis* for "way of living." Yellowstone Park, without wolves, suffered from dysbiosis.

Likewise, when we pull the microbes out of our food system, it negatively affects all the organisms along the food chain that depend on that food system. The microbiome of the soil affects the nutritional value of the plant, and the nutritional value of the plant affects the health of the animal that eats it. The plants you feed an animal, whether it is a steer or your son, also affect the composition of his gut microbiome. And it is the gut microbiome that manages the nutrition that animals get from their food.

Animals have evolved to eat particular foods in association with the bacteria that help them digest those foods. When animals go off-script and eat something in large amounts that they don't have the microbial populations to digest, they can get sick. Beef cattle are a case in point.

Most cattle in Iowa, and really all over the USA, are raised on grass on small or midsize farms and ranches, but then they are sold to a few large feedlots where they are fed corn and silage until they reach slaughter weight. You might think cattle are digesting grass and corn, but actually, they just chew the food and deliver it to the resident microbes in their guts, and it's the microbes (and their by-products) that the cattle consume for nutrition.

Let's get more granular. Cows have a four-chambered stomach. The first chamber, the rumen, is a 50-gallon fermentation tank with a quadrillion bacteria, protists, archaea, and fungi living inside, one of the densest microbial habitats on Earth. Cows chew the grass and deliver cellulose to fiber-digesting bacteria. The bacteria break down the cellulose into digestible sugars. Cud is simply a second round of chewing that increases surface area for the microbes to work on (the smaller the particles, the more surface area, and the more surface area, the busier the microbes can get). Protists in the rumen eat the bacteria and produce by-products like hydrogen, and archaea, which literally reside on the protists, break down the hydrogen into methane gas, which the cow emits from one end or the other. Fungi live in the rumen, too, but their role is unknown. Twenty percent of US agriculture's contribution to total US greenhouse emissions comes from farting and burping livestock.*

The mass of grass in the rumen is refined further as it goes through the honeycomb of gut chamber number two, the reticulum (source of the choicest edible tripe). The mass then passes into the third chamber, where all the liquid is squeezed out and the microbes that traveled with the grass mass are digested by the cow (a major source of protein), and finally into chamber four, the true stomach, where gastric juices finish the job.

Here's where the trouble starts. Cows like corn, and we like to feed them corn because it makes them fat and their meat super creamy and tender, but corn is mostly starch, which the cow's grass-digesting bacteria don't care for. As the level of corn in the diet increases, the cow's ability to digest grass decreases. This is because the cow's microbial community has to shift from microbes that primarily degrade cellulose to microbes that primarily degrade starch. By feeding corn, the farmer selects against the cow's cellulose-eating bacteria. If farmers switch to feeding corn to their cows too quickly, acid, a by-product of rapidly increasing numbers of starch-degrading bacteria, builds up, causing acidic blood, which leads to decreased milk production, reduced body fat, and diarrhea. And a sick cow is not an ideal food source for us.

All animal microbiomes are communities with populations of different types in different numbers. Setting aside an actual infection, whether an animal gets sick or not depends not on the type of bacteria present but on the ratio of one type to another. It's just like our lab on Yellowstone: too few wolves, too many elk. When population numbers change, the dynamic of the entire community is altered, and that changes the ecosystem, even causing dysbiosis: too few wolves, too many elk, eroding stream banks.

---

*Microbes in dairy cows release up to 264 gallons of methane into the atmosphere each day (a fact which delights the authors of ExxonMobil's policy blog, *Perspectives*). Other big livestock-producing countries have similar statistics. Sixteen percent of greenhouse gas emissions in Australia come from agriculture, 66 percent of which is methane produced by microbes and burped by livestock. And it's volatile. Methane gas released by dairy cows blew up a cow barn in Germany in 2014. Australian researchers have discovered that some sheep burp less than others; they are looking to use these sheep to produce a low-methane breed called "green livestock."

I took to exploring campus during the 2-hour break between class and lab. I wandered into the Dodge Physical Fitness Center, thinking I'd take advantage of the program: cardio kickboxing, vinyasa yoga, Zumba. Maybe I'd try kayaking, or scuba diving, which was held in the pool, or pick up handball or squash. The weather was warming up, and I had great ambitions to use the facilities, but I never did. It quickly became apparent that despite my growing confidence as a student, there were still hours and hours of reading to be accomplished, and on top of that, the lab reports took about the same mental energy and time as filling out my tax returns. Every week. It turned out that the only facilities I got to know well were the coffee stations in the student lounge and the library.

The library reading rooms are marvelous, like a very large English gentleman's study, fragrant with floor cleaner. But it's the stacks, where the library's two million books are stored, that are the core of the building. Literally. The stacks are in the center of the building, 15 floors of them open 24 hours, with reading rooms and offices around the periphery. It is, as Ben Ratliff wrote in the *New York Times*, like the library's inner organ. When you are in the stacks, there is no connection to the outside world. Two to three levels of stacks equal one floor of the library, so the ceilings are low and the shelves are close. They are intensely dense, claustrophobic spaces. When you enter this warren of books, smelling of dust and fungi and paper and ink, you have to turn on the lights as needed. The lights are on timers, which, if you have a tendency to get lost in reading, can be a real problem. Once when I was a student at Barnard, I don't know how long I had been reading, but at some point, I heard a click and the lights went out, and I panicked because I felt like I was suffocating, like I was trapped in a sarcophagus with walls of thick books, the only sound a murmur of dry, shuffling pages.

But not all students feel the same way.

On the Columbia students' blog, there is an infamous guide to having sex in the stacks. (But not on the library roof, one intrepid student advised. It's gravelly.) Written in 2005 by Chris Beam, he argued the library is a place where men can prowl with the added benefit of seeming innocent because they were also studying. "Many women see Butler as a restraining order waiting to happen," wrote Beam, but the "momentary correlations between academic achievement and sexual prowess make us all Casanovas." One commenter pointed out

that Butler sex between the bookshelves is nothing compared to Union Theo-
logical Seminary sex (UTS is an independent—and the nation's oldest—
Christian seminary located in Columbia University). "One floor above the
library lined with stained glass windows is a seldom visited hallway. *O Deus
Ego Amo Te*." Stuff like this makes me wish I were 20 again.

Sex in the stacks is illegal, as it is in any public place, but it's about on par
with sex in airplane bathrooms. Some flight attendants will just tell you to cut
it out; others will file a report and you could be slapped with an indecent exposure charge. As one stewardess wrote in response to a survey about sex in plane
lavatories, "It's certainly not high on the list of places I'd want to have sex,
especially given that the lavatory is cramped, smelly, and the liquid on the floor
isn't water. . . . But to each their own." Airplane bathrooms are also, according
to *Health* magazine, one of the 12 germiest places you'll encounter in an average day (along with yoga mats and hotel room television remotes).

Before our first official—as in graded—lab, I snuggled up on a stained
couch in the student lounge to read an assigned scientific paper, a stunningly
boring examination of a wee fossilized anklebone that revealed the common
ancestry of an ancient hippo and a contemporary whale. I read the same sentence multiple times before realizing it, and finally gave up, hoping that all
would be illuminated in lab. Lab was held in an extension of Schermerhorn
Hall, in a room nearly impossible to find. I wasn't the only person lost. I ended
up tagging along with a freshman who received instructions over her phone
from mission control, aka a girlfriend waiting for her in the classroom. Enter
the main building, go through the double doors, walk down a flight of stairs,
go through more double doors, down a winding hall clogged with obstacles like
students struggling to push dinosaur bones on rolling stainless steel tables with
gimpy wheels, down some stairs, then through a long hall in what feels like the
basement, and into our large room with dark windows and shared tables. I
never knew what floor I was on. Or even if I was aboveground.

Kyle stood at a podium and recommended we find a lab partner, and I
looked eagerly to my right and left, but both students had turned to the other
student next to them. It was rather like being the last picked on a team. I ended
up doing all my lab work on my own. This proved very difficult because the
reports required knowledge of computer programs I'd never mastered. I'd

never made graphs or charts. I'd never computed anything on my computer. I just used it for typing. Indeed, sometimes I'll hit something on the menu bar at the top of a page I am typing, and then I can't get rid of the weird formatting unless I shut everything down and take a walk around the block. Honestly, I even find making footnotes unnerving. When Kyle announced we had to create tables, I blanched.

Our first lab report was about phylogeny, the evolutionary development of species. We were given five Legos with different qualities, like color and number of studs, and had to figure out an order of their evolution from one type to the next. To do this, we had to apply the law of parsimony, or Occam's razor, a principle of problem solving which says that among competing hypotheses, the one with the fewest assumptions is the one that should be selected. We named the Legos with Latin-sounding names (mine were Lego, Lengtha, Colore, Triangula, *Triangula perfecta*) and made matrices illustrating shared characters. Are Lego and Lengtha the same color? Do they share a triangular feature? Since I had no idea how to create tables, I ended up waving Kyle over to my desk so many times that I had to apologize to the entire class for monopolizing him. But the kids were very patient and didn't complain even though I pretty much dominated the lab by frequently demanding "Which button do I push?" A young woman who sat in front of me with a tattoo behind her ear that said "Ridicurious" made a graph of her data in like 5 seconds and was turning it around and observing it in three dimensions on her screen like a CT scan of a tumor. I ended up drawing my graph on paper and taking a picture of it with my phone.

Once home I spent hours on the lab report. I was obsessed with capturing the language, the use of terminology as shorthand ("The data presented here illustrate a hypothesis for a phylogeny with homologous characters among the ingroup"), and the consecutive piling on of data, like layers of sediment in an ancient ocean, that backs up every statement, that takes us up to what we know now. Reports were due on Sunday night at 12:00 a.m. Every minute a lab report was late counted against our final grade, though we did have a 10-hour credit that we could use.

I couldn't imagine turning in a lab report past midnight, mainly because I am in a deep sleep by that time, but also, I was just way too conscientious, although I did screw up an assignment once, in Professor Rubenstein's section before the holidays. It was the only time I hadn't completed the reading material before class. Look, I'd had an aging-parent emergency, and I was way too worn out and put out and bummed out to struggle with a chapter on genetic regulation and so skipped it, okay? But of course—and this is why we still have nightmares about not being prepared for a class*—that day he decided we would break into small working groups to review our understanding of the reading material. There were only four of us in my group, an Israeli boy who was clearly sailing through the class, a level-headed African American woman, and a petite blond who watched me through narrowed eyes with that special kind of hatred that daughters reserve for their mothers. I listened to them discuss how the *lac* operon protein works. "It's simple," said the young man, leaning back in his seat and reciting from memory, as if he were verifying his address. "In the absence of lactose, the *lac* repressor binds to DNA at the operator site." It was mortifying. All I could do was shrug and smile apologetically, same way I do when the computer guy fixes my printer by plugging it in.

As the labs progressed, we got into Michaelis-Menten kinetics, a model of how enzymes work, and what it takes to make a biochemical reaction happen. Professor Boelman, who explained it, mentioned a woman was codeveloper of the model. While a small step toward correcting generations of discrimination against women in science, no amount of recognition could make this concept any more palatable. The Michaelis constant is the substrate concentration at which the reaction velocity is 50 percent of the maximum velocity achieved by the system at maximum substrate concentrations. What the hell does that mean? I never understood the notion for more than a couple of seconds at a time. It was like capturing a snowflake only to have it melt right away. And of course, I was technically too handicapped to translate the model into graph form on my computer. I felt myself becoming increasingly frustrated and curmudgeonly in lab, and for the first time, I sensed my behavior was like that of

* Psychiatrists have said that dreams about screwing up at school reflect a kind of meritocracy anxiety based on the belief that you can earn advancement and the fear you might fail to do so. It's common among people who studied to get where they are at in life. "I'll bet Prince Charles wouldn't have such dreams," Dr. Melvin Lansky, a professor of psychiatry at UCLA told the *New York Times*. "His place doesn't depend on merit."

an elderly person who gets angry at his phone because he can't seem to dial the correct number. Kyle was infinitely patient, though I could tell he was resigned to me being the problem child.

I did, however, become proficient at taking screenshots, and my lab file had hundreds of them. I took pictures of the graphs every step along the way, just in case something spontaneously erased. It was totally paranoid, like running though the combination on the lock you use on your gym locker over and over again to make sure you've got the combination right. And I got better at the lab reports in some respects. I think I was pretty good at titles, like "Let Bivalves Be Bivalves: How environment influenced morphological evolution in *Ensis directus, Mercenaria mercenaria, Mytilus edulis,* and *Spisula solidissima,*" but I never mastered graph making and had to redo them constantly because I'd forget to change the date range or I'd screw up the error bars, or whatever. "First calculate the means and standard errors," we were told. But when I looked up the standard error equation, it didn't even look like math, for God's sake.

At the beginning of March, I told Jonathan I had a test coming up, and he cancelled his other tutoring sessions. "I am very serious about tests," he said. We hunkered down. He quizzed me relentlessly for hours. I made detailed flashcards, which I practiced on the subway and while cooking dinner. I'd repeat the answers back to Jonathan exactly, and then he'd make me say them in different words. I carried around sticky notes with the definitions of *entropy* and *enthalpy, reduction* and *oxidation, cofactors* and *coenzymes* in all my pockets, and a week later, when I pulled my blue jeans out of the dryer, it took me a minute to remember why they were sprinkled with shreds of yellow paper.

My mind was so preoccupied with biology that after Professor Boelman explained water physics, I looked at my husband's martini glass differently. "Look, Kevin," I said to him more than once, "the reason it is bulging at the top is because of cohesion!" When a friend of ours asked how our dinner conversation was going now that I was back in school, the implication being that I might have become a tad single-minded, my husband gave me his "see what I mean?" look.

Professor Boelman's exam turned out to be as short and harsh as a rejection letter, heavy on cofactors and substrates and how energy influences chemistry.

Even though she kept saying we're not going into the weeds, the test was totally weedy. There was so much jargon it would have made a consultant's head spin. But I survived it, thanks, in great part, to Jonathan.

"I take how my students do on tests personally," he said, and he let me make him a cup of green tea.

The second session, which began right after our test (no break, no homework reprieve), was taught by Matt Palmer, my advisor, an urban and peri-urban ecologist, which means he studies things like what kinds of plants and insects and birds move in when a city baseball field is allowed to rewild. By the end of the school year, he'd merged with Professor Naeem's lab, now the Naeem-Palmer Lab Group. Matt was like a Broadway star, with a melodious voice, a big toothy smile, lots of goofy expressions, eyebrows all over the place. The kids loved him.

He taught mechanisms of evolution and ecology, which started with why and how populations contain so much physical variation, and he showed us a Far Side cartoon by Gary Larson that illustrated a bunch of Twinkie-shaped lemmings heading toward the water, one with a life preserver around its waist representing the mutant that survives.* On the first day of class he put up a picture of Darwin and asked, "Who's this?" and someone yelled, "Santa!" The examples of evolution in his lectures focused on charismatic organisms, like the finches that diversified into different species with different beaks to eat different kinds of seeds on the Galapagos Islands, and the population distribution of animals like cougars and deer.

But we didn't talk about how microbes might have contributed to the evolution of animal development, social behavior, and speciation, which they did. I had to look that up myself. Animals evolved with their microbes, and their microbes influenced their evolution. The four chambers of a cow's stomach probably evolved in tandem with the bacteria and protists that live inside it, and the cow's immune system evolved to protect the microbes it needs to digest food and destroy those bugs that are not in the family.

Microbes help animals adapt to and expand into new food niches, which is

*The notion that lemmings follow each other to death is a hoax created by the Walt Disney Company. In its Academy Award–winning *White Wilderness*, the filmmaker chucked the lemmings off a cliff. "Heigh-ho!" said Professor Palmer.

key to speciation. It doesn't make sense from nature's point of view to create a new way to eat from scratch when you can co-opt the skills of bacteria and let them do the work for you. In their billions of years on Earth, bacteria and archaea have figured out how to get energy from just about anything. Outsourcing this job allowed animals to vastly expand their repertoire of things they could eat and, in the process, evolve into the glorious variety of species on Earth today.

For example, bees may have diverged from their carnivorous wasp ancestors when they picked up new kinds of gut microbes that were able to extract energy from pollen. Up to 70,000 species of insects manage to eke out a living in a variety of pathetic niches because their gut microbes help them survive on nutrient-poor diets. Take sap-sucking aphids. Plant sap lacks essential amino acids, but bacteria living on waste in the aphid's cells produce amino acids for them. Most animals are herbivores, and every one—including most mammals, plant-eating lizards, birds, marsupials, and even herbivory dinosaurs—evolved symbiotic partnerships with plant-degrading microbes. Early mammals were probably carnivores, and even today, we lack the genes to encode enzymes that degrade plant cell walls. But with the help of our gut microbes, we can get nutrition from eating plants. Indeed, we may even be better off eating plants.

Even within our species, our microbes help us adapt to different food opportunities. When we feed our microbiome different foods, the microbial community structure changes. The consumption of grain is a case in point. When humans started farming and became more dependent on grain for their nutrition, the population ratios of gut microbes trended toward those that could manage carbohydrates. (The microbial population in their mouths changed, too. The microbes that cause gum disease increased. It happened again when sugary foods became widespread and fed a population surge in cavity-causing microbes.) Our gut microbiome adapted to the grain diet. Or you could look at it from the reverse. Our microbiome allowed for the transition from hunter-gatherers to farmers.

This is where the Paleo Diet argument, that human genes evolved before the onset of agriculture and therefore we should be eating the foods of hunter-gatherers, falls apart. Most of the genes in our bodies are genes within microbial cells, and most of those microbial cells are in our guts. The ratio of microbial types in the gut changes based on available food. If humans couldn't get nutrition from grain, there wouldn't be as many of us here. There are definitely problems with the modern diet, like obesity, but you can't blame that on grain. Those problems aren't more than 100 years old, and we've been living on wheat for about 10,000 years. Our genes don't program us to eat a certain diet. Our diet influences our microbiome's genes. Take the great panda. It is a bear, and bears are omnivores; they eat meat. But pandas can exclusively live on bamboo because they acquired the microbes to digest it.

We house 300 times more genes belonging to microbes than human genes. Some scientists describe our genome as enhanced by the genome of our microbiome, and that we are such exquisitely complex creatures because we're genetically supersized by the bugs in us. We actually don't have more coding genes than a mouse (of course we have lots of genes that do noncoding jobs), and yet we make organisms like Sappho and Woody Allen.

The range of influence microbes and their genetic material have within a body is dazzling. For example, they can affect development. When certain bacteria are present, they can initiate choanoflagellates, among the oldest animals, to change from being free-living unicellular creatures to a colony. Microbes can delineate and reinforce social groups. Termite bodies and hive walls are coated with a colony-specific type of intestinal bacteria. If you are the unlucky termite who wanders into the hive without wearing the right bacteria, the locals will identify you and take you out. Microbes emit specific volatile chemical signals that help animals communicate. Spraying, or territorial marking, is a kind of "chemical graffiti," as Ed Yong wrote in *I Contain Multitudes*, announcing all kinds of personal details. Bacteria in hyena urine identify the sprayer's sex, membership in a group, and reproductive state. Same goes for elephants, badgers, meerkats, and bats. That's something to remember next time you yank your dog away from a lamppost he's sniffing; actually, he's just checking his mail.

Symbiotic microbes—those that live on and in their host—have evolved to play a role in animal sex lives, too. Birds attract mates by spreading pungent oils made in their preen glands by bacteria. Grass grub beetles house bacteria near their vaginas that attract males (yes, they have a vagina). Certain bacteria in the male Mediterranean fruit fly increase their sexual competitiveness. In some flies, the presence of particular bacteria in the females makes them sexier to males.

Indeed, bacteria produce the aromas of sex and perspiration that turn *us* on or, in some cases, off. Bacterial diseases like syphilis are associated with distinct odors. In a study, women rated the armpit odor of men with gonorrhea, caused by the infection of the bacterium *Neisseria gonorrhoeae*, as nastier than the odor of men without the disease. It seems we have evolved to associate the smell of the bacteria with the, um, inappropriateness of a particular date. It makes me wonder if deodorants might be hiding diseases we would otherwise detect.

Symbiotic microbes even help animals get around their food's defenses and help the hunt. For example, bacteria in the drool of moose allow them to eat red fescue grass, which is host to an endophytic fungus that produces toxic alkaloids, and desert woodrats can feed on the poisonous creosote bush because their symbiotic bacteria help detoxify it. The deep-sea anglerfish has a feature like a car radio antenna that grows out of its head, and at the tip there is a blob of bioluminescent bacteria that, in the bible blackness of the abyssal zone, lures innocent fish in the dark.

All of these kinds of interactions between microbes and their hosts and the food opportunities in their environments have evolved over a long time. They are part of what makes a species a species. You can't accurately describe any particular animal without also describing its symbiotic microbes. In class, we learned that based on mitochondrial DNA, humans and chimps shared a common ancestor about 6 million years ago. Likewise, we house gut bacteria that share a common ancestor with the gut bacteria of chimps, and our last common microbial ancestor with chimps lived about 5.3 million years ago, after which we and our gut bacteria diverged from chimps. If we only had our gut microbes to identify us, you could still tell a chimp from a human. Microbes were, are, and always will be an integral part of what makes a particular animal adapted to its environment. To put it another way, microbes allow us to live where we do and, as a result, be who we are.

Before the days of cell phones, Columbia dorms had the ROLM, a phone system that allowed students to contact each other, like phones in a hotel, and parents to reach their children. It also had a voicemail system that allowed students to forward voicemails on to other students. One phone message from 1989 became famous enough to be featured on the NPR program *This American Life*. The student, Fred Schultz, had put a song from *The Little Mermaid* movie on his voice answering machine and requested the dialer leave a message for him and the Little Mermaid. His mother called, having done him a favor only if he promised to stay by the phone so she could deliver his requested information hassle-free, but, of course, being a college student, he had forgotten and taken off. Here's what he came home to. "Hi, Fred. You and The Little Mermaid can go *fuck* yourselves. I told you to stay near the phone. I can't find those books. You have other books here. It must be in La Jolla. Call me back. I'm not going to stay up all night for you. Goodbye." It became a campus-wide sensation, passed from student to student with message threads so long the campus voice-mail technology couldn't handle it and the messaging system for the whole of Columbia crashed.

The Columbia student body is notoriously communicative, and active. There's a light bulb joke: How many Columbia students does it take to change a light bulb? Seventy-six. One to change the light bulb, 50 to protest the light bulb's right to not change, and 25 to hold a counter-protest. (The Harvard version is, How many Harvard students does it take to change a light bulb? One. He holds the bulb and the world revolves around him.) During the Vietnam War, Columbia students occupied a campus building for a week, calling on the university to stop conducting weapons research for the war and to oppose the construction of a segregated gym. More than 700 students were arrested in 1977 when students blocked the appointment of Henry Kissinger to an endowed chair at the university. In 1982, the Coalition for a Free South Africa convinced the student senate to approve a motion to support divestiture. I remember that ragtag group, in jeans and boots and moral outrage, clogging up the entrance to the administrative building. At the height of the blockade, 1,000 students were sitting on the steps of Hamilton Hall. They received disciplinary threats; it went public and Jesse Jackson and Desmond Tutu weighed in with their support. Finally, the trustees caved, and in 1985 the university divested its assets in the apartheid South African government. The next year Congress passed the Comprehensive Anti-Apartheid Act that banned new investment in South Africa and prohibited sales to the police and military. F.W. de Klerk, the last president of the apartheid regime, told a *Chicago Tribune* reporter, "When the divestment movement began, I knew that apartheid had to end."

Today there are big and small protests on campus. Korean students in 2016 protested the Korean president, Park Geun-hye, as did hundreds of thousands of Koreans in South Korea. Students protested the wrestling team, which had been caught sharing racist, sexist messages. After failing to persuade the university and the police to pursue criminal charges against her alleged rapist, Emma Sulkowicz protested by lugging a 50-pound mattress like those supplied in dorms wherever she went on campus. She carried the mattress to her 2015 graduation ceremony, too. There were protests during my recent Columbia experience, as well. The installation of a Henry Moore sculpture drew the ire of quite a few students who thought it was ugly and an inappropriate fit for the university's predominantly neoclassical style (it was installed anyway and it's beautiful). Columbia Divest for Climate Justice, founded in 2012, is a long and slow boiler. Part of a national college movement, it calls for the divestment of college endowments in fossil fuels. In 2016 the Advisory Committee for Socially Responsible Investing recommended the university divest in companies that develop unconventional, high-cost resources like tar sands (a notoriously expensive, dirty, and ecologically devastating practice), which represent "denial by deed" of climate change science.* In April, I saw about 100 students setting up beds on the cold steps of Low Library, now the administration building, for an 8-day "Sleep-Out." "Let us stand in solidarity with the students at Columbia and NYU for demanding their schools divest from fossil fuels," tweeted Bernie Sanders.

Climate change deniers drive the scientists at Columbia crazy. Quite a few of our lab projects involved tracking changes in chemical reactions in plants due to changing climate (my associated lab report was titled "Hot and Heavy: Forest respiration in a changing climate"), and more than once we were shown

---

*The argument is there are adequate traditional sources of fossil fuels to meet a climate increase limit of 2 degrees centigrade, beyond which global warming would have serious consequences. To mine for fuel in excess of that limit, which explorations of exotic resources like the tar sands do, is to deny by deed the existence of climate change science.

the work of the Mauna Loa Observatory in Hawaii, which has been tracking monthly average carbon dioxide concentrations since 1960. While there are seasonal fluctuations in CO_2 in the atmosphere, the overall trend is up, up, up.

Evolution deniers drive them crazy, too. We learned the criticisms of evolutionary theory and the counterarguments that can be made: *Evolution is just a theory* (so is gravity, but I am not jumping off the Butler Library roof). *The organs of living creatures are too complex to have formed by a random process.* In the first semester, we learned evolution isn't random; it's a process where random mutations are favored or selected over time. But what I learned in Professor Palmer's section was evolution doesn't act on individuals. It acts on *groups* of individuals, or rather, on gene pools. Here's the example he gave. A hog bred for its meat is not, as an individual, fit to survive, because it gets slaughtered and made into sausage. However, because the genes of the hog breed make it fat and delicious, the population is fit to survive, because the hog has genes that are selected by their environment, which is our breakfast appetite.

The same is true of microbes. Evolution doesn't act on them as individuals, but as groups. Likewise, evolution doesn't act on individual microbes in us; it acts on microbial communities on a population scale. As a result, a microbiome follows the same patterns of community ecology as the members of Yellowstone Park: succession, competitive exclusion, reaction to disturbances—all concepts we learned in class.

Imagine ancient Yellowstone Park when it was an erupting volcano. How we get our microbes is much the same as how the first colonizers of the cooling caldera got theirs: in successional waves. We are mostly born sterile, like a volcanic island, and we acquire our first microbiome from our mothers and our surroundings, just like plants. A child born naturally is exposed to a population of microbes in Mom's vagina as well as some of Mom's gut microbes via her stool. This is the first wave of microbial succession in a human host. Over time, the microbiome shifts from one that was determined by the population at birth to the next successional wave, a microbiome populated by microbes from other sources, like food and environment.

The competitive exclusion principle says no two species can occupy the exact same niche indefinitely, not when resources like food and space are limited. In class, Professor Palmer described two species of barnacles living on coastal rocks, one that can only survive when covered with water, so they live near the bottom of the rocks and are covered even at low tide, and another species that can survive out of water for short periods of time and so found a niche higher up the rocks where they can live without competition. The two barnacle groups eat the same stuff, but they evolved into separate species that allowed them to divvy up the territory so they didn't have to compete for food. Our microbiome is similar. In a mature gut, every niche is filled, and competition for food among the microbes—as long as that food source is consistent—is minimized. In a healthy gut, all those niches are filled by microbes that have evolved to maintain the well-being of their ecosystem, which is you. A balanced and diverse ecosystem, where there aren't too many wolves or too many elk, is a stable ecosystem.

But like a forest after a forest fire, if the microbial community is disturbed by severe illness or, more likely these days, by antibiotics, the community becomes destabilized. "Most microbes in us are good," said mycologist and plant pathologist James White. "That is why antibiotics are so wrong." Taking antibiotics is a little like setting a forest fire to kill a specific weed. It disrupts the entire microbial community because a course of broad-spectrum antibiotics kills both the pathogen you are being treated for and a bunch of other bacteria, as well. That affects the population ratios of the microbes that live all over your body, not just in your gut, which in turns affects the functions those bacteria perform.

Because there are just so many spots at the table that is you, when niches open up from death by antibiotic, someone will take the available seat. Microbes with resistance to the antibiotic may multiply, or fugitive species of microbes might get a foothold. Maybe percentages of a species are killed, causing a ratio change in the population, with some species (and, by extension, their functions) decreasing in number and others increasing. Which means the seats at your table that used to be filled with microbes that performed certain duties, like a plumber and an electrician, are now filled with microbes that might do different things, like a draper and an exterminator.

Except when a particular species of bacteria gets into a place where they are not supposed to be, a bacterium is not bad or good. Indeed, the whole notion of "good" bacteria and "bad" bacteria is misguided. There is only a spectrum of community structures that range from normal bacterial communities that have achieved an ecological balance that you benefit from by feeling good and abnormal bacterial communities that make you feel sick, that are dysbiotic. It's like the elk. They aren't bad in themselves; they're just bad in the wrong numbers for the physical environment.

But luckily, a healthy ecosystem wants to bounce back and return to its original structure. The microbes that live in and on us have evolved in and on us, and our immune system recognizes them as friends. There likely are stores of these bacteria tucked in hiding places that reassemble their population when the antibiotic is gone, which is why you may have diarrhea or a skin rash for a while after a round of penicillin but usually get back to normal before long. Just as in the recovered ecosystem in Yellowstone, once the wolves were reintroduced, those organisms that had an evolutionary niche to live in—the wolves and elks and beavers and fish and willows—increased or decreased in population numbers until ecological balance was attained.

By April, the snow in New York City had crumbled and melted, and the students started to forget their hats and scarves on their chairs, and I began thinking of my body like Yellowstone Park. Maintaining my health, suggest the authors of the paper "The Application of Ecological Theory toward an Understanding of the Human Microbiome," should be more like park management: habitat restoration, promotion of native species, and the targeted removal of invasive species. Studying ecology was the window by which I came to understand my body in a deeper way. There's a name for this kind of realization. It's what the scholar and environmental activist Joanna Macy calls "the greening of the self."

Our Microbiomes

A s soon as the weather warmed and the days were longer and the furry buds on the magnolia tree in the Wien Courtyard were fat as little mice, the girls began to show up at class in dresses with bare, goose-pimpled legs and the boys went sockless, their knobby ankles red with cold. In anticipation of the democratic primaries, students prepared a spring "Dorm Storm for Bernie." Spring changed the tempo of the campus. No longer hidden in their hoods, the students yelled across the campus to their friends and lingered in the sunlight, perched on the cold cement benches around the central lawn like birds waiting for crumbs, their jackets open to reveal T-shirts that said "Feel the Bern" or images of Senator Bernie Sanders as a Jedi master. My son was into him, too. That spring he put aside his Pugs Not Drugs T-shirt and wore one that said "Bernie Sanders: Eat the Rich."

From what I could tell by their T-shirts, most of the student body was liberal, but I knew there were alternative views, some extreme. The online hangout for procrastinating students, Bored@Butler (as in Butler Library), had its share of "Make America White Again" types. As one student wrote, "Where would I be without my racist online message board for Columbians by Columbians?" But as one commenter elsewhere pointed out, judging Columbia by Bored@Butler was like judging a university by what's written on the bathroom walls. Bored@Butler went offline a few months later, having declined in general use due to the "explosive growth of alt-right online culture." Bored@Baker, Dartmouth's decade-old chat group, was shut down in 2015 because it became sodden with odious speech, too, even sporting a rape guide. Anonymous forums are always popular places to host offensive opinions. Indeed, a demographic analysis available at WikiCU pointed out that Bored@Butler visitors "happen to be fucking idiots."

Hateful idiocy is terrible. But there's a certain kind of idiot humor in college that I remember finding insanely funny when I was young, and that seems to blossom in spring. "Schmasterpieces of Western Lit," a take on Columbia's core curriculum Masterpieces of Western Literature, is a student video depicting a drunk girl sitting in a bathroom stall trying to retell the story of *The Odyssey*. "And Odysseus is like 'Yo I'm smart that's my thing I got this' and so while the Cyclops is asleep Odysseus stabs him in his eye and he's like only got one and the Cyclops is like 'Odysseus you asshole you blinded me.'"

Actually, this still makes me laugh.

On these raw spring days, walking across the campus to the subway, I often felt nostalgic about being college-aged, about being thoroughly entertained by inane jokes, about being unequivocally sure of my rightness. I guess loneliness and anxiety led me to think about springs past—the glassy, fragile ice on the pond near my childhood home in the morning, the mushroomy smell of the damp dirt road we lived on, the pale green hue that lingered like a question over the woods. And then, as predicted by Constantine Sedikides, a professor of social and personality psychology at the University of Southampton in the UK who realized reliving good feelings restores hope, I'd start to feel better, even elated, and I'd reconnect to the purpose of being back in school: to understand how I am connected to the budding magnolia tree, to the kids I pass on the college walk.

Our body is a biome, a major habitat for trillions of microbial cells that have found niches all over us, from our scalps and guts to between our toes. What we do with our bodies and the bodies of our children—how we are born, what we eat, what chemicals we are exposed to—affects whether we acquire the right microbes at the right time and whether we can maintain the mature communities of microbes we need.

We get our initial microbiome from Mom. It's our first birthday gift, wrote mycologist Nicholas Money in *The Amoeba in the Room*, and it is composed of the microbes swarming in the vagina. The vaginome, as it is sometimes called, is dominated by a few species in the genus *Lactobacillus*—one of which is the kind you find in yogurt—that produce lactic acid. Lactic acid keeps the vagina acidic, an important barrier against unwelcome bacteria like those that cause chlamydia and gonorrhea, and viruses like herpes and possibly Zika. Lactobacilli also produce hydrogen peroxide, which you use to clean a cut, but in the vagina, it beats back the growth of group B streptococcus, a leading cause of infant mortality. The ideal pH of a human vagina is between 3.8 and 4.5, about the acidity of an apricot. If the acidity goes beyond those parameters, it can undermine the population of pH-sensitive bacteria, and then microbes that can hack the new pH environment like the fungus *Candida albicans* may increase. That's what vaginosis is.*

The composition of the vaginome can change with sexual partners or can differ between different races of women, but once we get pregnant, all our vaginas move into the same lactobacillus-dominated state, and for good reason. In a recent study, the presence of these bacteria lowered the risk of preterm birth.

The current thinking is the amniotic sac is sterile, so before delivery, the baby has no microbiome (though there is evidence of microbial communities in amniotic fluid, the placenta, fallopian tubes, the cervical canal, and uterus, suggesting a nonsterile environment). But when a baby is born naturally, he is coated with vaginal microbes. The vagina is "like a glove," wrote Dr. Martin J. Blaser in *Missing Microbes* (he's the Moses of gut microbes; his research has led the way in understanding the importance of gut microbial diversity), molding around every surface, crease, and crinkle, giving the baby a good dose of those lactobacilli all over his body and some in his mouth, as well.

The baby is also exposed to Mom's fecal microbes, or should be. When I arrived at the hospital in labor with my first child, the nurses told me my doctor had ordered an enema. I suspected my doctor wanted it done so I didn't soil him, but I'd heard that having an enema made labor more painful, and since I was in serious pain, I said no way. It turns out that was a good idea because when my baby was born, a small amount of stool inoculated her. If all goes well, your baby will be born facedown (when the baby presents faceup, it's called back labor). It

*Not much is known about penis microbes, though circumcision is known to alter the, yes, penisome. One can't help but wonder what benefits are lost due to circumcision, because a foreskin provides a nice airless hiding place for microbes, not unlike the tissues that once surrounded a corn kernel. The inside of the foreskin has a layer of mucus in which immune cells reside. When activated by the presence of an invader, these immune cells produce antigens, which in turn induce immune responses. Less mucus, less immune cells; less immune cells, fewer antigens; fewer antigens, less immune responses.

helps the baby get that first mouthful, and not all mouthfuls are equal. The microbiome a baby inherits from Mom's stool reflects the food she eats, and the food she eats reflects the farming methods used and the soil it was grown in.

The vernix, that cheesy stuff covering a newborn, is often washed off in the hospital. That's too bad because it contains useful proteins that suppress specific pathogenic bacteria. Indeed, the World Health Organization advises delaying baby's first bath for 24 hours. A newborn's immune system is underdeveloped, which is convenient because it allows for colonization by Mom's bacteria, but it's also why young children are prone to infection. But Mom's bacteria waste no time colonizing her baby's skin and guts right away. We inherit our microbiome from Mom and the environment, just as a seed does.*

But when a baby is born by Cesarean section (and according to the Centers for Disease Control and Prevention, over 32 percent of all births in the USA in 2014 were C-sections), his first dose of bacteria is skin bacteria, which are floating around in the room because humans shed skin bacteria constantly. The skin bacteria colonize the baby inside and out, in essence competing with lactobacillus. Researchers have found that babies born by C-section are at greater risk of developing asthma, allergies, obesity, and celiac disease, among other conditions. It's not all bad news. A 2017 paper in the journal *Nature* suggested that as long as a mother had labor before her C-section, the vaginal microbiota still made their way from Mom to infant.

The microbes introduced at birth determine the community structure of a child's gut microbiome, maybe for her entire life. There's an order to the colonization of microbes in our guts. Baby's first colonizers are supposed to be predominantly lactobacilli, the microbes picked up in the vaginal canal. Lactobacillus sets up a healthy human gut with positive influence over digestion and immune functions. If other species of bacteria are baby's first colonizers, the baby's microbiome sets up differently, maybe harmfully. Diseases like asthma are thought to occur in conjunction with a gut microbiome that is maladapted— that isn't growing the way nature intended, that's become an island of weeds.†

---

*Curiously, the microbial communities from an infant gut are more closely related to those of plant roots, and therefore soil microbiota, than those associated with adults.

†Additionally, a variety of gut microbes synthesize vitamin K. Newborns who do not acquire the gut microbes they need, or they don't travel to the baby's gut quickly enough, are at risk of developing vitamin K deficiency, which can lead to an inability of blood to clot. All newborns in the UK and USA are now given vitamin K shots at birth.

Similarly, a baby who is put on antibiotics at birth faces the same problem of establishing those important first microbes because while antibiotics are extremely effective at killing targeted microbes, they also can cause a lot of collateral damage. Most babies born in hospitals in the USA get some antibiotics in the form of drops in their eyes, to ward off a type of pink eye caused by chlamydia or gonorrhea (in case Mom has it). This practice has its risks. A study of mice showed that eyes are loaded with bacteria that seem to play a role in stimulating the immune system of the eye.

Babies treated with antibiotics before their first birthday are more likely to be later diagnosed with asthma. In a study tracking babies' microbiota from birth to 3 years old, those at risk for asthma were found to harbor lower numbers of some gut bacteria, and in a study of mice infected with at-risk babies' stool, researchers found the introduction of a "healthy" microbiota reduced their airway inflammation.

To mitigate this, some midwives are swabbing C-section babies all over and in their mouths with Mom's vaginal bacteria. There is a risk that pathogenic bacteria like group B streptococcus or venereal diseases could be spread to the baby, but if the swab carries no pathogenic bacteria, the practice could help those children develop a gut microbiota similar to natural birth babies. So far it seems swabbed babies only receive partial microbiota restoration—maybe due to the fact that the bacteria are transferred via gauze or the diversity of microbes transferred from Mom is reduced by the antibiotic treatments she underwent as part of her C-section. Anyway, the long-term effectiveness of the practice has not been demonstrated in humans, but it does seem to work in mice.

Breast milk is the first food of human babies worldwide. Or it should be. There are lots of good reasons to breastfeed, both for mother and child, like bonding, the perfect infant nutrition, and decreased risk of breast cancer. But breast milk does something else: It feeds the microbes in your baby's gut. It "engineers a baby's ecosystem," wrote Ed Yong in *I Contain Multitudes*. Early breast milk is a delivery system for *Bifidobacterium longum infantis,* one of the earliest colonizers of the infant gut and, along with lactobacillus, the dominant microbe in the guts of breastfed infants. By a woman's eighth month of pregnancy, *B. infantis* are in and around her nipples secreting acids and antibiotics that repel pathogenic microbes like *Staphylococcus aureus,* which grow in milk ducts. When the baby sucks, he picks up the bacteria. In this sense, breast milk is a probiotic (a source of bacteria) and a prebiotic (food for bacteria). Breast milk is composed of lactose, fats, and particular oligosaccharides that babies can't digest and don't get from any other food but breast milk, but *B. infantis* love. Your breast milk feeds the microbes that protect your baby from, well, microbes; infants who aren't stuffed with *B. infantis* may suffer more GI problems. And it's dosed. In tandem with the stabilizing of baby's microbiome, Mom's supply of oligosaccharides for *B. infantis* declines and her supply of lactose and fat increases to accommodate her baby's evolving needs. Bottle-feeding, in contrast, is a blunt nutrition instrument. It just doesn't feed the baby the right ratio of nutrients at the right time. Indeed, the World Health Organization recommends 6 months of exclusive breastfeeding. Twenty-five years ago, a decade before the term *microbiome* was coined, I was told as a young mother to breastfeed for a minimum of 3 months because breast milk shared immunities that compensated for my baby's developing immune system. That recommendation aligns with what the new microbiology is telling us.

Once breastfeeding declines and vegetables, cereals, and animal products like chicken and eggs and cheese are introduced into a child's diet, the lactobacilli and other bacteria cede dominance—but don't disappear—to secondary-succession microbes introduced through her food, and by the time she is ready to enter kindergarten, a mature community has stabilized. Microbiome imaging suggests the composition of the gut microbiome is not segmented into species-specific groups, with one type living to the east or to the west of another, but rather, the community is mixed, like cocktail nuts—and that implies a certain stability. But that community can be destabilized at any time in our lives, when we take antibiotics, for example, or alter our diet drastically (like going vegan, or maybe even eating foods with too many sulfites, which are antimicrobials that can harm bacteria), but it doesn't change easily. While population ratios can shift in as little as 5 days, it's hard for a new species to move in unless a niche opens up. A mature gut microbiome is like a mature ecosystem. It tends toward stability.

Like plants, we actively construct our microbiome. Plants choose the microbes they need and we do, too, when we choose whether to eat a kale salad or a Twinkie. While most breastfed babies have a similar gut microbiome, once we start making choices, in what we feed them and what drugs we give them, we determine the specifics of our children's resident microbes. Who's in there and in what numbers influences your child's nutrition, his immune system, his chemistry, even his mood. It's one thing to realize I've been ignorant my whole life about how my food choices affect the microbes that I depend on. It's quite another to realize the food I fed my child affected the microbes he depends on, too.

We love to hear about centenarians who smoke or eat only bacon and cookies. I think it allows us a reprieve from what most of us know, that minimally processed foods, vegetable-rich diets, and constrained use of alcohol are better for your health—that you are what you eat. My dad is 91 at the time of this writing. He is tough and creased like an old baseball mitt, a World War II vet who shovels snow, and weeds his garden, and unloads shot in the general direction of varmints with his heavy rusty shotgun. He eats mainly homegrown vegetables, some organic meat, and lots of small wild seafood from the Atlantic like mackerel, whiting, and squid. He is Italian American and eats pasta quite a few times during the week and salad every day (at lunch, not dinner, as it is hard for him to digest pasta and raw stuff at night). Dad drinks his own rough wine and eats bread with meals, some cheese, and hardly any sweets except fruit. I've been following his eating patterns for years because I'd like to be shooting at woodchucks in my garden at 91, too.

To that end, every Friday morning I meet him on Arthur Avenue in the Bronx, a street bustling with Italian markets, to observe his shopping and pick up recipes. We always meet for a coffee at Tino's Delicatessen. The owner, Giancarlo, is short and stout. He has decorated his delicatessen with long shared tables, autographed headshots of celebrities, and a very large and very expensive marble bust of Marcus Aurelius, often adorned with ties and sunglasses. The deli dishes out meatball subs to the local firemen and sells specialty products from Italy to Italophiles from the tristate area. While my dad and I are sipping our cappuccinos, Giancarlo will come around and say, "You see I gotta the burrata from Italy? It's very good. I ate two last night," and he closes his eyes and shrugs, and I think he looks kind of sad, and then I realize he looks sad because he's not eating the burrata right now.

My dad is proud of his diet and very proud of his poops, which he frequently describes in much the same way Giancarlo describes his burrata, so when I suggested he send a stool sample to the personal metagenomics company uBiota for a microbiota scan, he was thrilled. It was like he was getting an opportunity to take a test he knew he was going to ace. And it was easy, too, because Edward uses the facilities promptly after his cappuccino at Tino's. uBiota sent me two kits to try ($80 each)—chocolate brown boxes decorated with bright green ovals—and all we had to do was swirl a cotton swab on some soiled bathroom tissue, drop it in a baggie marked biohazard, and put the box in the mail. Edward's sample was prodigious. I filled out our online questionnaires about our weights and diets: Were we low carb? Halal? Paleo? Ovo-lacto? And our restrictions: Shellfish? Gluten? Tree nuts? And details about the sample shape. I actually had to guess my Dad's, but considering his braggadocio, I assumed he was a Bristol Stool Scale type 3 or 4—like a sausage, the best kind. And then we waited.

Prior to 2013, the only way you could get a genetic analysis of your own microbiome was to join a study. uBiota and companies like it (uBiome is another) have two kinds of customers. The first are people like my dad, who pay to have the microbial genome of their stool or some other part of their body analyzed because it is interesting or they are trying to self-diagnose a chronic problem or they want to track the effects of a dietary change. The most basic service tells you what kinds of microbes are present. The other customer is the scientific community. (Many of these companies skew toward the clinician and don't sell do-it-yourself kits, like Microbiome Insights and Second Genome Solutions.) Your de-identified microbiome data, as well as lots of other people's, is made available as a research database. The American Gut Project is all about this crowd-sourcing angle. It allows you to compare your gut microbes to those of the guts of thousands of other people in the USA including food writer Michael Pollan's. American Gut's open-source data is included in the Earth Microbiome Project, a massive, ambitious effort to characterize the global microbial genome. It's one mother of a database.

For under a hundred dollars, you can find out who is living in your gut. But it turns out that's like trying to determine who lives in Times Square based on a snapshot of who's on the street at the time. The population ratios of your gut microbiome can change depending on your recent diet or if you've just gotten off antibiotics. Usually that change isn't likely to be permanent; it's just a reflection of the environment—your diet and health and immune function—selecting for certain populations at that time. This is true of microbial communities in general. During the Deepwater Horizon oil spill in the Gulf of Mexico, the microbial community shifted to one dominated by microbes capable of hydrocarbon degradation because there was a lot of oil to eat. But an hour before the spill started, there were not as many hydrocarbon-degrading microbes in the Gulf. Additionally, fecal samples aren't intestinal samples, though one bowel movement can liberate a third of the bacteria in your colon. "A fecal sample is like looking for a key under a lamppost," said the microbiologist Moselio Schaechter. "You find things because that's where the light is."

After a few weeks, I got an email from uBiota with our Personal Metagenomics Reports attached. These turned out to be colorful pie charts that allowed us to see the percentages of the bacterial phyla, family, order, genus, and species found in our samples. I showed my dad the charts on my computer, and after clicking through them for a while, he said, "So . . . what?" It was, as the journalist Tina Hesman Saey described, like "looking at a stranger's yearbook; you can put names with faces, but have no idea what the people are like or why you should care."

Dad's and my guts turned out to be pretty similar. We are, or were the day of our sampling, dominated by bacteria in the phylum Firmicutes and, to a lesser degree, the Bacteroidetes, of which there are a whopping number of species. Firmicutes are associated with a high-calorie diet, meaning lots of animal products, and Bacteroidetes metabolize carbohydrates. Pretty typical, pointed out Kael Fischer, the cofounder of uBiota. Our samples showed similar microbes all the way to the species level, most of which didn't even have names, though we contained them in different ratios. I had thought for sure my dad's diet would show the presence of microbes that were more like those in the microbiome of, I don't know, an Amazonian tribesman than the prime minister of France.

Western microbiomes tend to be less diverse than those in nonindustrialized populations. This may be a result of our antibiotic use in everything from medicines to animal husbandry to hand wipes or our reduced exposure to bacteria from wild sources like soil (a good reason why we should let our toddlers play in the mud). Likewise, increasingly antiseptic and engineered foods—baby food is virtually sterile—has led to a reduction of the diversity of microbes in our body, and according to James White, "in the long term that may have negative indications." The molecular anthropologist Christina Warinner investigated 1,500-year-old Celtic feces and found it contained five species of a fiber-fermenting bacterial genus, *Treponema*, that we don't have anymore. That's in part because we collect microbes from our environment by eating, as well as from soil and other organisms, so the more sterile our environment, the less types of microbes turn up in our guts. Just like the soil story and the plant story, a less diverse microbiome is less able to deal with environmental stresses from pathogens, antibiotic use, or dietary changes or opportunities.

But it's quite hard to increase one's microbial diversity. For example, we can't acquire the kind of diversity enjoyed by the Hadza of Tanzania by taking supplements. We'd have to live the hunter-gatherer lifestyle. Microbes considered healthy in us, like bifidobacteria, aren't even present in the Hadza. Each one of us—Hadza, you and me—are all individual ecosystems shaped by the ecosystem in which we live.

As we neared our final lab reports, Kyle warned us that lab partners should be very careful not to copy from each other. That wasn't a problem for me because I didn't even have a lab partner, but considering how difficult the material was at times, I wouldn't be surprised if, under duress from exam week or a breakup or any of the myriad things that can throw a student off-kilter, it might be tempting to ask someone to share her hypothesis on the effect of global dimming on species distribution. One Columbia lab TA (who prefers to remain anonymous) told me about two sorority sisters who were lab partners. "One produced beautiful reports. The other? Couldn't care less." But toward the end of the semester, they turned in identical lab reports. Technically, TAs are supposed to report cheating directly to the Office of Student Conduct and Community Standards, where a first-time offense may earn an F for the class and repeat offenders could get expelled. But instead, he went to the professor. "We didn't accuse them, we didn't say plagiarize or cheat. But they freaked out and the good student took the fall," said the TA. "I don't know how they thought they wouldn't get caught. My guess is it's pretty rare that someone gets kicked out of Columbia for cheating. It's too bad, because ultimately, it underserves the student and reinforces privilege."

In a world that is rife with appropriation made easy by the internet, some people may get mixed up as to what cheating is, though you'd think one's moral compass would be enough.* In 2013 Columbia and Barnard had to deal with a wide-ranging cheating scandal. In a big lecture class where students were allowed to self-grade, weekly quizzes were consistently hitting the 90s. The cheating was ironic because the class was known for its easy A. (Students take the occasional no-brainer course mainly to boost their GPA, otherwise known as Rocks for Jocks and Physics for Future Presidents. And one can assume courses with titles like Twilight: The Texts and the Fandom just can't be that hard.)

The book *Cheating in College: Why Students Do It and What Educators Can Do about It* points out that cheating habits among college students probably develop long before they get into college, and that most college-bound students are exposed to significant cheating cultures during their high school years. Not me. In the 1970s I went to a hippie school where behavior lacking in integrity could easily be the subject of a school-wide shaming-cum-reconciliation assembly. The price for cheating was double. You wouldn't just get in trouble for doing it; you'd also have to endure the expression of the community's feelings of betrayal.

I sent in my last lab report well before midnight. I made sure I had a good title: "Size Matters: The effect of random chance on the genetic structure of model populations." ("If you have a good title," advised my tutor Jonathan, "you get more citations.") In my acknowledgments I thanked Kyle for his patience, especially with the graphics. I liked writing the acknowledgments. It wasn't the norm at Columbia to applaud our professors after a lecture, any more than applauding when your plane lands, yet an expression of gratitude seemed due. Which is why, by the second semester, I broke from the pack and made a point of thanking my professors after their presentations.

In my last class before the final exam, Professor Palmer explained the concepts of communities. The individualistic concept of Henry Gleason says that communities are aggregations of species co-occurring at one place where each

---

*I've heard that frats and sororities at Columbia keep files of old exams from every class their members have attended, which (if true) straddles the fault line between solidarity and ethics.

species responds individually to the environment. The holistic concept by Frederic Clements says that communities function as an integrated unit, a superorganism. "What conditions would favor one concept over the other?" he asked.

I knew an answer. In the human gut, microbial communities function as an integrated unit that benefits the host. But those microbial species have the potential to respond individually to environmental changes, specifically to the food and drugs we ingest. We are superorganisms, but depending on what kind of environment we provide them, one or more of our symbiotic communities are totally capable of going rogue. I hoped the question would be on the final exam.

The microbes on and in us, the diversity of species, the community structure of those microbes, and the ratio of types are a reflection of our lifestyles. They shift with our choices. There are differences between the microbial community structures of a vegan and a meatatarian.

But putting aside those differences, the microbial body map is pretty consistent among industrialized populations. The initial limited microbiome of a baby gives way to a greater diversity of microbe types, picked up from food and liquid, from the environment, pets, wherever. These microbes assort and find a home in every nook and cranny of your body, from the space between your eyebrows to your sinuses and lungs, from your mouth to your gut to your urethra—a cornucopia of different species with different lifestyles.

In general, our mouths house over a thousand species of bacteria, as well as archaea, viruses, sometimes fungi, and amoebae, some of which may pay their way by performing duties like regulation of blood pressure, but most of their identities and functions are mysterious. It's what the geneticist Rob DeSalle calls an "open system." Mouth microbes are being introduced all the time as a result of kissing—eighty million bacteria can be shared in a 10-second French kiss, among them, spit-eating bacteria—ingestion and breathing. The bacteria responsible for morning breath hate oxygen, so they live in our mouths in smaller numbers during the day when air is going in and out but proliferate at night because we usually breathe through our noses when we sleep. There's a regional component to oral microbiomes. You are exposed to species that are symbionts of foods and landscapes particular to your geography, and through kissing and spit, species that are particular to your family and lovers.

Teeth have several habitats, and different sides of a tooth can harbor different communities. Some microbes like enamel; others prefer soft tissue. When these communities are disrupted, maybe by eating too many candy canes, pathogenic bacteria can increase, form biofilms, and their by-products can cause illnesses. Bacteria that ferment sugar produce acids as a by-product, and the acids dissolve tooth minerals, aka dental decay. Periodontal disease is caused by inflamed gum tissue that has been irritated by the presence of plaque, or bacterial biofilm, on the tooth.

Heading south, there aren't any permanent resident microbes in the esophagus that scientists know of yet, but there are in our stomach, a highly acidic environment that dissolves the food we eat. Scavengers and carnivores have more gastric acid than herbivores because the risk of encountering a pathogen is higher in a meal of roadkill than in an ear of corn. Depending on your gastric acidity, most pathogenic bacteria don't make it past our stomachs: The low pH environment either kills them or we throw them up. (We have taste buds in our guts. They are identical to those in our mouths, but smaller and live in clumps; among other duties, they make us vomit when we eat noxious foods.)

Our stomach is home to a variety of species, most famously, *Helicobacter pylori*. In 2005, the physician Barry Marshall won a Nobel for his work proving *H. pylori* caused gastritis and ulcers, which he accomplished in part by drinking a test tube full of the stuff. His work unleashed a pogrom against the bacteria. "The Only Good *Helicobacter pylori* Is a Dead *Helicobacter pylori*" was the title of one paper. It turns out to be an ancient human symbiont: the intensely poked and prodded mummified remains of Ötzi the Iceman, a 5,300-year-old body pulled from a thawing glacier in the Alps, revealed evidence of *H. pylori*. At the beginning of the 20th century nearly everyone housed *H. pylori* in their stomachs (it is transmitted from person to person by saliva and by fecal contamination of food and water). The bacteria are in less than half the human population today. But according to research done by Dr. Martin J. Blaser, *H. pylori* is not all bad. People with resident *H. pylori* are less prone to allergy and asthma. When the stomach is empty, *H. pylori* produce gastric hormones that are essential for making ghrelin, the appetite hormone, and when the

stomach is full, they stop producing the hormone. It could be that the war on
H. pylori is screwing up these signals, which may be contributing to the epidemic of obesity. Dr. Blaser has suggested that inoculating children with
H. pylori could shield them from asthma and obesity, and then, once they are
old enough to fall into the demographic likely to get an ulcer, the *H. pylori* bacteria could be eradicated. That's a revolutionary idea for our kill-the-bug model
of medicine. It's more like park management.

The acidity of your stomach shapes the diversity and composition of
microbial communities that make their way further down the GI tract, where
the small intestine comes next. Most of the nutrition we get from our food is
processed and delivered to the bloodstream in the small intestine. It is a lower-
acid environment than the stomach, about the acidity of a tomato, and as a
result, the diversity of microbes increases 10,000 times. Scientists aren't sure
what these microbes are up to. Enzymes coded by our own genes break down
most of the foods digested at this point in the gut, like animal products and
processed foods like white bread. The leftovers, what aren't digested in the
small intestine, like plant fiber, head straight for your colon, where they are
digested by microbes using *their* enzymes.

Here, things slow down. The large intestine, the colon, is the Mississippi of
the body—five feet of the big muddy, where the pH is mild, like an ear of sweet
corn, and the living is easy. Lots of microbes find a home here—100 billion
cells, perhaps 2,000 different species, altogether about 3 pounds of them competing and collaborating for the plant fiber that made it south. These are old
friends that perform two primary roles. They interact with our immune system
in support of human health, and they pump out molecules that are key to the
production of a virtual pharmacy of chemicals we need.

Gut microbes have evolved a symbiotic relationship with our immune system. They protect us and we protect them. One way that gut microbes in the
colon provide immune services is by crowding out potential invaders, much the
same way as resident bacteria fill all the niches on the leaves of a plant. A "normal" gut microbial community, one that is rich in bacteria that ferment fiber
we don't digest on our own (also called resistant starch), inhibits the growth of
pathogens like the potentially fatal *Clostridium difficile* by taking up all the
available space. *C. difficile* can make spores that resist antibiotics. If multiple

rounds of antibiotics knock out too many residents of your "normal" microbial community, then *C. difficile* can get a foothold. The best treatment, in that case, is to undergo a fecal transplant, which has a 90 percent cure rate. In this procedure, a dose of microbes from a healthy microbiome is squirted into the *C. difficile* sufferer. No one knows exactly what combination of organisms in the stool is helping. The principle at work is *C. difficile* is an invasive weed that can be outcompeted by the injection of competitors that don't do you harm. However, a small German study in 2017 showed that *C. difficile* patients who were administered fecal transplants where all the bacteria were filtered out also recovered from the infection. No bacteria were administered, but tinier things—viruses and secondary metabolites that got through the filter—were, so maybe bacteriophages, those viruses that prey on bacteria, are the real reason people get well from this procedure. With *C. difficile* out of the way, there's real estate available in the gut that can be recolonized by our evolutionary symbionts.

The fecal transplant approach to restoring gut flora has been around for a long time. The British zoologist Jane Goodall noticed that chimps ate fellow chimps' feces when suffering from diarrhea. A 4th century Chinese physician gave patients with severe diarrhea what would come to be known as "yellow soup," made from the feces of a healthy person. There is anecdotal evidence fecal transplants can help patients with a range of disorders. A batch of businesses has sprung up to take advantage of the fecal transplant miracle (and its lack of FDA oversight). OpenBiome, for example, will freeze a patient's stools in a microbial bank so they can repopulate their guts after antibiotic therapy. They're personalized probiotics. Or they can provide someone in need of a transplant with stool from well-vetted anonymous donors with code names like "Winnie the Poo," "Dumpledore," "Albutt Einstein," and "Vladamir Pootin." My Dad, confident in his bowel movements, suggested I sell his on the internet to supplement my writing income.

As much as 60 percent of immune system tissue is located around our intestines, particularly at the cecum, the entrance of our large intestine. That's also where most of the microbes are. Their presence primes the immune system, similar to how endophytic fungi prime the immune system of plants. The whole human GI tract is huge, about 30 feet long. "The gut is the biggest area of interaction between the environment [meaning, organisms distinct from ours] and the body," said Dr. Fischer of uBiota, "so the immune system is all over it, checking it out." If all is well, our bodies accept these microbes, having learned when we were infants to distinguish between the microbes with whom we have an evolutionary kinship and the microbes that cause disease. The presence of our symbiotic microbes may help the immune system stay focused on the real enemies. Autoimmune diseases can happen when that discretion fails and the immune system mistakes something okay to be in the body for something that is harmful.

The immune system isn't just about attacking pathogens. It's also concerned with regulating inflammation—the body's immune response to disease—and gut microbiota play a role. The colon wall is only one-cell thick. These cells produce mucus that lines the inside of the colon. Mucus helps make the wall impenetrable and helps, well, slide things along. Living on the mucus layer are bacteria that ferment undigested starch. Those bacteria signal the cells as to how much mucus to make. If the mucus-lining bacteria are disturbed or underfed and the chemical signals they provide that tell the cells what to do decline or disappear, the cells won't produce enough mucus to maintain the firewall between the contents of the colon and the bloodstream, and the colon leaks. There are all kinds of bacteria in the colon, and some of them produce toxic molecules. When those molecules leak into the bloodstream, the immune system kicks into action and starts an inflammatory response. Ongoing inflammation can lead to chronic diseases.

Alterations in intestinal microbiota have been linked to many diseases. It's correlative, but perturbations in intestinal microbial ecology have been associated with celiac disease, pediatric multiple sclerosis, rheumatoid arthritis, obesity, diabetes, malnutrition, and inflammatory bowel disease (IBD), which includes ulcerative colitis and Crohn's disease, all of which have strong known links to the physiology of the immune system. Parkinson's and chronic fatigue syndrome have also been associated with an impaired microbiota. Environmental factors like smoking and genetics play a role in diseases like IBD, which may be why fecal transplantation isn't very effective at curing this disorder.* But genes alone don't explain the dramatic increase in IBD since World War II,

---

*Though research continues, albeit slowly. The FDA considers the feces, in this capacity, to be a drug, and so it must be tested under the Investigational New Drug program.

notably in industrialized countries. Environmental disruptions like a course of antibiotics or a high-sugar diet play a role, as they undermine the healthy assembly of bacteria that protect us from chronic inflammation.

In a healthy GI system, the by-products of fiber-fermenting gut microbes participate in feedback loops with our immune system, helping to lower levels of inflammation to maintain a comfortable environment for themselves, and we benefit from their comfort. ("Remember," said James White, "there's no moral relativity in the microbe-host relationship.") Gut microbes throughout our GI system break down proteins and carbs and sugars for themselves, but in so doing, they increase the bioavailability of the trace minerals we need. Those in our colon assist in stimulating water and salt absorption. They also tell the pancreas and liver to get ready to work, regulate the levels of fats and sugar in the bloodstream, and synthesize a number of important vitamins, like vitamins K and B_{12}, and essential fatty acids necessary for energy and helping the body absorb dietary minerals. The colon microbiome is the Big Pharma of the body, pumping out needed molecules for the smooth operation of virtually every part of our body, even our brain (which, coincidently, weighs about the same amount as the microbes in our guts).

"Friends with brain benefits," as the neuroscientist John Cryan describes them, are microbes that help produce neurotransmitters like serotonin, most of which is made in the gut. This may explain why folks that suffer from depression and ADHD also tend to suffer from GI problems—dysbiosis of the gut flora may cause both gut and behavioral problems. Even disorders like schizophrenia and autism may be linked to an impaired gut microbiota in early childhood, when the bulk of brain development happens. In animal studies, mice raised without their normal microbiota lost their preference to interact socially, but then when their microbes were replaced by means of a fecal transplant, they became social again. And without their microbes, when placed in stressful circumstances like having to get off a high step, they produced twice as much stress hormone as mice with normal microbiomes. Gut microbes may be involved in what's known as the gut–brain axis, the communication between the central nervous system and the mesh of neurons that governs the functions of the GI tract. If microbial production of neurotransmitters is interrupted, that could influence brain function. There is a lot of work to be done on this—really, all aspects of the human microbiome—but it conjures up a future where one day microbial probiotics might be prescribed for mental disease.

So important are these microbes to our overall health—to our ability to obtain nutrition, modify our immune system, and maybe even regulate our mood—that we've evolved a little organ to house spares. Some scientists think the appendix is a savings bank for bacteria. "A peaceful alley," wrote Rob Dunn in *The Wild Life of Our Bodies*, where bacteria can grow securely and be on hand in case the GI tract needs recolonization. The bacteria stored in the appendix may be introduced at birth when baby gets her first gulp of Mom's stool. It's like Mom opened a bank account for baby, and the deposit was the symbionts of a mature human. When a disease like cholera sweeps through or we take broad-spectrum antibiotics, wiping out much of our bacterial residents, the bacteria in the appendix are in the wings, available to move in. People who have had an appendectomy probably pick up their familiar microbes from food and the environment, but it takes a while, and in the meantime, they could be vulnerable to an adverse infection. Scientists at Winthrop-University Hospital in Mineola, New York, found that appendectomy patients with a history of *C. difficile* were more than twice as likely to have recurring infections.*

There is no practical separation between my microbes and me. Just as in soil and plants, microbes upload nutrition into their host in exchange for housing, and they maintain a stable home environment. We need them and they need us. Microbes are a determining factor in our health, and they may even affect how we feel about ourselves, whether we are optimistic or melancholy. It almost seems like microbes are the puppet masters and we are the puppets. But since our choices, like what we eat, affect their world, I suppose we are only puppets if the choices we make are founded in ignorance. It's the well-informed puppet that can pull the strings.

---

*What happens if you have a colonoscopy? Same thing. It takes a few weeks for your colon microbiota to return to normal, so even if you haven't had an appendectomy, it's probably a bad idea for a New Yorker to kayak the Amazon River right after.

Germophobia and
Microbiomania

W e took our final exam on a sunny afternoon in early May. All the classroom windows were open, and the magnolia tree was in full and royal bloom. The students smelled sweet, like spring water. The last question was for extra credit: Write a poem about ecology.

> There once was a girl from Korea
> Who ate nothing but sweets for a ye-a
> But those gut bugs she fed
> Replaced others instead
> And now she's got diarrhea.

I never picked up my last test, and to this day I wonder how I did and if Professor Palmer was amused by my limerick. But I didn't have to wait long for my final grade. My professors were quite efficient at getting them out, which is a mercy, because waiting for grades is a particular kind of anxiety. Actually, I checked into one of the online student chat groups that I occasionally turned to for comradery, and everyone was weighing in on what it's like to wait for a final grade. Students described it as akin to waiting for the results of an STD test, or waiting to hear about graduate school, or waiting on the results of your girl-friend's pregnancy test when a certain latex sheath didn't perform its assigned duty. For me, it was like waiting to get the results of a mammogram, which pretty much explains why I only lurked in these chat rooms.

Columbia has a tradition of calling out teachers who dawdle, and the student blog has lists of "The Damned." It's quite long and drenched with frustration. "Update: Still no update." I checked CourseWorks about five times a day from the last day of class until I received my grade a few weeks later, but I acted casual when I told my husband over dinner that I'd gotten an A-. I couldn't tell if the joy that spread across his face was because I had done well or because he was utterly relieved the semester was over and I would no longer be his disagreeable, unavailable, and unkempt roommate and would actually pay attention to him again.

I was relieved, too. Relieved to be done, relieved to have my life back, relieved I'd performed adequately. But most of all, I was relieved to no longer be totally at sea when I read a news story about Legionnaires' disease or the label on a probiotic yogurt. These weren't just words with vague meanings anymore. It was like learning to swim. Suddenly, the water became accessible.

I didn't attend my Barnard graduation back in the 1980s. I was just ready to move on with my life, and my parents didn't seem to mind. For some, graduation is a big deal. It's the end of life as most students have known it for quite a few years, and people respond in different ways. Some kids party, others are in a panic about their future prospects. Some avoid the problem of graduating altogether by staying in school. There are the ones that just want to move on and don't care about graduation, and the ones who try to finish their bucket list, like exploring the labyrinthine tunnel system under Columbia's campus, and the emotional ones, who say "This is the last time I will ever see you!" and it's often true. I had a friend at Barnard, a beautiful Californian surfer who hooked up with a wealthy friend of her father's. She lived in his dark brownstone on Sutton Place filled with statuettes and Sotheby's catalogues, but she always seemed to be alone. The final day of classes she wore a full-length mink coat with nothing on underneath but a bikini, which caused a minor tsunami in our Italian seminar. It was also the last time I saw her.

This time around, I paid attention. Columbia has a commencement week calendar that counts down to the second, like the countdown clock at NASA or the countdown of shopping days left before Christmas. The graduation action starts with the graduation fair. At the back of the Columbia bookstore, half a dozen tables sported the essentials: personalized college announcements (basic, essential, or deluxe); diploma frames, which a saleslady described in the soft persuasive tones of a coffin vendor, "Honey walnut molding with a soft sheen and hints of gloss finish"; and class rings with names like the Galaxie and the Lady Legend. Class rings show affiliation, like a biker jacket. Supposedly Mark Antony gave a kind of group ring to his Praetorian Guard. The modern practice dates back to 1835, when West Point handed them out, and eventually other schools took up the practice. I wanted one, even though they are butt-ugly. But I wavered because it was kind of phony—my group dispersed decades ago—even though the nice lady at the ring table, with a knowing nod, said, "No problem. We back-date them all the time." And of course, graduation regalia, the cap and gown, were for sale, by degree and height. They are pale blue with black trim and white appliqué crowns: $57 for a BA, and $64 for an MA robe, because it has wings.* The petite lady who worked this table was a pro with the kind of good-humored, hard-bitten efficiency of an ER nurse or a toll operator. "Master, bachelor, PhD, height?" she barked as the future graduates stepped forward. I asked if I could take a picture of the gowns. "Knock yourself out," she said.

Before the Civil War, Columbia students wore black caps and gowns to class every day (wool for bachelor's degree candidates, silk for master's, and, ooh, velvet for PhDs). After 1894, the gowns only made an appearance during commencement, and in the late 20th century there was a gown update to the light blue number worn today. The current model is made from 100 percent postconsumer plastic bottles. Professors wear the cap and gown to commencement, too. My friend Guido, an Italian who teaches architecture at Cooper Union in New York City, wears regalia from the Istituto Universitario di Architettura di Venezia, where he graduated. His cap looks like a fur-covered gondola.

"These gowns are not rentals, but souvenirs—they are yours to keep, without

---

*Columbia's history is its brand, and this is true of all the Ivy League schools. Columbia was King's College for 30 years until the American Revolution in 1776, when, unsurprisingly, the name became politically incorrect and was changed to Columbia. The crown symbol refers to the old King's College; they are all over campus if you look for them, on the gates to Butler Library, carved into the foundation stone of Hamilton Hall. The colors, blue and white, are derived from the insignia colors of the upperclassmen's Philolexian Society, one of the oldest collegiate literary societies in the United States, known for partying, bad poetry contests, and debating such issues as "Is there an objectively best sandwich?" and the Peithologian Society, founded a few years later so freshmen could have something to join.

the trouble of pre-orders or returns," intones the Commencement page, a souvenir that you probably will never wear again unless you go on to teach. (Barnard students get their robes for free.) My daughter, Carson, graduated from the University of Toronto. "Oh, graduation robes," she wrote me when I asked. "They're way too hot, never fit right, and are made from the most god-awful itchy material. . . . I'm convinced it's the university's last stab at trying to break students before their big moment. We rented my robes (they were like 90 bucks), and the hoods identify your graduate level, arts versus science, and honors. My hood was lined with white rabbit fur (undergraduate degree with honors in arts). I didn't get to keep any of it, which means some new undergrad sweated through it during their big day shortly after I took mine off . . . kinda gross."

Of course, University of Toronto professionally cleans the regalia between graduates since shared cloth is a way to spread disease. When Carson was around five, she was in a children's theater group where the instructor let the kids try on costumes and figure out characters and spontaneously put on a play. I always thought it was more of an experiment for the teacher, who was very interested in how the kids self-assorted into roles. But my daughter loved it. Unfortunately, the bin of costumes was also home to a poxvirus, *Molluscum contagiosum*, or water warts. Tiny pinkish blisters erupted all over her torso. The treatment our dermatologist applied was worse than the disease. She used curettage, or scraping the spots off with a razor, which stung and was bloody and rather horrifying. Needless to say, my daughter didn't go back to theater school.

That experience was enough to make any mom a germophobe. We have a love-hate relationship with microbes. On the one hand, low-grade germophobia is rampant. It has been since the early 20th century when advertisers for products like soap and Listerine tapped into the social anxieties of people moving from farms to factories to offices, according to Katherine Ashenburg, author of *The Dirt on Clean*. Marketing, in combination with public awareness of germ theory, is how bad breath became a medical condition, halitosis.

On the other hand, new discoveries about the important roles the microbiome plays in human health has created what the biologist Jonathan Eisen calls the "overselling of the microbiome," and the subsequent opportunism of diet book doctors and skin-care companies seeking to utilize sparse science for hefty gains. At first, I thought the truth of our relationship with microbes lay somewhere between germophobia and Dr. Eisen's "microbiomania." But that's not accurate. Our relationship with microbes includes it all, from the most devastating illnesses to the most miraculous-seeming cures. What needs to be teased out is the truth buried in the hype: the *actual* amount of flesh-eating bacteria cases and what difference probiotics *really* make. Once I had a little microbial literacy under my belt, I could apply what I'd learned about microbes to my daily interactions with these extremes.

Germophobia is a pathological fear of contamination by germs, that is, everything microscopic that could possibly cause illness. The physician William Alexander Hammond coined the term *mysophobia* (which means the same thing) in 1879, from a case of excessive handwashing, now thought to be an indication of obsessive-compulsive disorder. It's a pretty famous condition: sufferers include Howard Hughes, Saddam Hussein, and Donald Trump. ("The extravagantly rich . . . are our most notorious germphobics, people made uncomfortable by the thought of shaking a stranger's hand," pointed out Mary Roach in the *New York Times*.) Sufferers relieve their anxiety with ritual cleaning, even though they know their obsession is crazy. Antidepressants seem to help, possibly because they control the anxiety. I know when I was a child and our household was tense with marital conflict, I had my compulsions. I used to check the fridge door was closed over and over, many times a day. I don't do that anymore, but just in case I was latent OCD, I took a quiz, Are you a germophobe? The questions were along the lines of "Does it disgust you to share bottled water with someone?" (A little.) and "Do you wash your hands as soon as come into the house?" (Yes. I ride the subway.) My results: 39 percent germophobe. "You aren't a germophobe, but you're not dirty either. This means you are aware of what is gross but you don't let it take over your life."

But high-wattage germophobes have terrible problems navigating a social world. And the news doesn't make things easier. I have a Google Alert for "bacteria," and every day I receive notifications like "Millions of Bacteria Lurking in Our Water Pipes!" "Scientists Identify Potentially Harmful Bacteria in Espresso Machines!" "Harmful Bacteria Thrive on Cookies!" That's just the tip of the iceberg. "E-Coli Bacteria Found in Ice Cubes!" "Is Your ATM Dispensing Bacteria?" "The Bacteria in Your Beer Pong Cups Is Off the Charts. CHEERS!" Beware your dollar bills, I'm warned, and your loofahs, your cell phones, your unwashed bras! Your hot tub could give you hot tub lung, a bacterial infection spread by steam. *Prevention* magazine warns of the bugs living on your kitchen faucet, the garbage disposal, even the welcome mat. Most handled means most germy; bathroom hand towels are covered in bacteria, mainly because we often skimp on the handwashing (if you care about eradicating germs, wash your hands long enough to sing the Happy Birthday song twice). Ground-floor elevator buttons, touch screens, the coffeepot handle at work . . . if you believe the headlines, dangerous germs contaminate everything.

Airplanes and public bathrooms are treacherous traps in which the germophobe can fall, though toilet seats are actually not the worst sources of contamination. They are designed to keep as much bacteria off you as possible, and anyway, many of the bacteria in our bodies are anaerobes. They don't survive long once exposed to air. (Though there are many, like the horrid *Clostridium difficile*, that *can* survive.) The 5-second rule—that food lying on the floor for more than 5 seconds is unsafe to eat—is just plain silly. Bacteria can get on that food in a nanosecond, and anyway, contamination of dropped food has more to do with the porous nature of the food than how long it was on the floor. A study found that a gummy bear dropped on the floor doesn't pick up near as many bacteria as a chunk of watermelon. The very fact that microbes are everywhere is why we needn't worry much about them. Most of the bacteria we encounter in a bathroom or on the subway are harmless skin microbes, which we are shedding all the time.

If everything that is you disappeared, mused the biologist Clair Folsome, "what would remain would be a ghostly image, the skin outlined by a shimmer of bacteria, fungi . . . and various other microbial inhabitants." Our skin, about 2 square meters of it, is our largest barrier to pathogens, our first defense against disease. Skin insulates, keeps water in (and out), regulates temperature, generates sensations, and synthesizes vitamin D when exposed to sunlight. And the microbes that live on our skin are part of that shield. Like the other microbial ecologies I learned about (soil, plants, and the human gut), a well-developed skin microbiome with diverse critters maintains an ecological balance that keeps pathogenic bacteria from getting a foothold.* Likewise, an imbalanced skin microbiome, or one reflecting the inflammatory responses of a screwed-up gut microbiome, may play a role in psoriasis, atopic dermatitis, and eczema.

---

*A small study showed that the dominance of certain skin bacteria, and the chemicals those microbes produce, determined whether you are one of those people mosquitoes like or not.

Our skin provides a lot of niches. For bacterial species, that falls into three environments: the moist (armpit, groin, and toe webs), the oily (head, neck, and trunk), and the dry (forearms and legs). So far scientists have found that a few bacterial phyla typically dominate each region. Fungal communities are present as well, mostly as harmless freeloaders, but they can become problematic if the balance of power within the microbial community is compromised. Too few bacteria to counter them and fungi run amuck. That's what a yeast infection is.

Microbial niches are so specific that we have different fungi that live between different toes and different bacterial species on the right versus the left hand, even different fingertips carry different communities. And those site-specific communities are similar from one person to the next. As Julie Segre of the National Human Genome Research Institute explained in the online magazine *Motherboard,* "The bend of your left elbow is most similar to the bend of your right elbow. But the bend of your elbow is more similar to the bend of *my* elbow, than your underarm."

There are 1,000 to 10,000 species and millions of microbial cells on a bit of skin the size of a letter key on my computer keyboard, and even more in humid areas. When you bathe, you wash some of these organisms off, including the transient ones that are usually the trouble-making pathogens (and it doesn't matter if the water is hot or cold, the result is the same). When you hug, you share them, which is why people who live together have similar skin microbiomes. In fact, a study found that two roller derby teams shared their skin microbiota within their team, but after a game, after all the shoving and body checks were over, the two teams' skin microbiota had homogenized. We are connected to each other by skin microbes, and the measure of that connection is determined by how much time we spend together. Microbes put a new spin on the notion of quality time.

After birth, Mom's vaginal microbes colonize the baby's skin. This initial skin microbiome is composed of the same critters that migrate into the gut and set up the baby's gut microbiome. A baby's skin microbiome subsequently diversifies with environmental exposure, which includes the outdoors, cuddling, and curious household pets. Come puberty, adolescent hormones increase the secretion of oil, which can lead to a proliferation of oil-eating bacteria, which live in oil glands embedded in hair follicles. They ferment the oil and produce fatty acids that irritate the follicle and cause an inflammatory response—a zit. Acne is overabundant *Propionibacterium acnes,* a microbe that exists on unblemished skin as well. To deal with acne, five million prescriptions for oral antibiotics are written each year in the USA for an average duration of 300 days. As a result, the species is becoming resistant to common oral and topical antibiotics, as are other skin bacteria. Both *P. acnes* and *Staphylococcus aureus,* which is associated with skin disease, are resistant to erythromycin, a popular antibiotic used for treating acne. The future of treating acne lies in developing medicines that kill *P. acnes* but don't lead to microbial resistance.

Armpit microbes face similar genocides every morning with a smear of Ban. Antiperspirants and deodorants (which often include ethanol or other antimicrobials) alter the microbial ecosystem of the pit, changing both the type and quantity of bacteria. But for that matter, so does washing with soap. Armpits are stinky because bacteria living in your pits break down sweat into thioalcohols, which smell like sulfur and onions, and feet are smelly for the same reason, just with different bacteria producing a slightly different-smelling by-product (and athlete's foot produces yet another smell, one caused by fungi). Without the bacteria, sweat doesn't smell. We all have unique armpit microbiomes, but some people's microbiomes are home to more of the thioalcohol-producing bacteria than others. Scientists at the University of California, San Diego, did a smelly pit transplant, where a twin with really bad B.O. washed with antibacterial soap and then received pit bacteria from his sweeter-smelling brother; the B.O. disappeared.

I decided to try the Mother Dirt AO+ Mist ($49) as an alternative to my regular underarm product. Sweat produces ammonia, and AO+ Mist contains live cultured ammonia-oxidizing bacteria (AOB) common in soil and untreated water. AOB—that is, bacteria that "eat" ammonia—reproduce slowly, dividing every 10 hours or so, and while they may well have been a part of our ancestral skin microbiome, with frequent washing, they don't get a chance to make a stand. Mother Dirt's concept is if AOBs can outcompete the sweat-eating bacteria that produce thioalcohols, you'll have sweeter-smelling armpits.

I used the mist for a couple of months and reduced my bathing to every third day, which I realize probably undermined the experiment, but I did experience a small decrease in underarm odor. Clearly it was a very unscientific trial, plus I hadn't gone to the gym for months, so I was not the most obvious candidate. I don't know for sure the AOBs were even alive the whole time as I kept forgetting to put the mist back in the refrigerator in order to keep the bacteria stable. A friend in my mushroom club recommended I use it on my "lady parts," which seemed effective, but rather chilly first thing in the morning. Vagina odor is a reflection of resident bacteria. If the bacterial colony becomes dominated by anaerobes, which tend to produce foul-smelling by-products, then there's an indication the pH has changed, which might be the result of an out-of-whack microbial population. This can happen for many reasons, including having sex: semen is basic—it has a pH of around eight, about as acidic as an egg white—which temporarily decreases normal vaginal acidity.

Diet can influence the makeup of skin microbial communities. Bacteria that produce funky underarm smells feed on lipids, so if you don't want to use deodorants or spritz ammonia-oxidizing bacteria on yourself, you can always start eating more vegetables, because fast food and meat make your armpits smell worse. Indeed, microbes produce most of our body smells.* The selling of hygiene is tricky because it's not just a matter of dirty or clean. Disinfecting your body is a kind of ethnic cleansing, with no one in particular slated to move into the evacuated space. But without a doubt, someone will move in. There will always be microbes colonizing your skin, no matter how many times you wash your hands.

The cosmetics industry has embraced microbiomania with enthusiasm. Products from the BioEsse Probiotic Skin Care System claim to have an extract taken from "beneficial" bacteria that "nurtures friendly microbes on your skin." Elizabeth Arden SUPERSTART Skin Renewal Booster ($67) claims to boost your skin's natural defenses with lactobacillus. TULA Exfoliating

---

*We can tolerate the smell of our own farts because we are familiar with their microbial composition. But our disgust with other people's farts may be our brains' way of saying stay away from potential disease, as farts can contain pathogenic bacteria. Mothers, by the way, perceive their infants' excrement (aka their microbes) as less offensive than the excrement of another baby.

Treatment Mask ($54) contains "signature probiotic technology," Tata Harper Purifying Mask ($65) is "probiotic-powered," and Rodial Super Acids products (sleep serum: $110) use "beneficial biotilys probiotic technology." Note these descriptions don't really say the cream contains probiotic cultures—live bacteria— because they don't, though the specifics of what they do contain is proprietary information. Dead bacterial culture could confer some biological benefits, as long as what is beneficial about it—a protein in the membrane of the cell, for example—has nothing to do with it being alive. But if so, "the rational scientist would purify this apparently very potent molecule or protein and develop it as the next big thing," pointed out the microbiologist Philip Strandwitz.

Probiotics may make one's complexion look better. There's some research that suggests a 3-month program of an oral *Lactobacillus plantarum* supplement improves skin elasticity and hydration. And they may make us appear more sexy and affectionate, too. A study at MIT found that when mice were fed a probiotic that originated in human breast milk, their fur grew luxurious, the males grew large testicles, and the females started pumping a fertility protein and the hormone oxytocin, otherwise known as the love hormone or the cuddle chemical. (Cuddling is associated with mothering and social bonding, which, in the process, transmits microbes.)

Just to get the terminology straight, a probiotic is alive. It is bacteria that are reintroduced into a location where they belong. Probiotics include species that are permanent residents in your guts and transients, microbes that are just passing through but play important roles in your gut ecology during their travels. Theoretically, a probiotic supplement (they are primarily species of the bacteria *Lactobacillus* and *Bifidobacteria* but also fungi like *Saccharomyces boulardii*) returns permanent residents to the neighborhood, or increases the supply of beneficial transients while you are taking them. The point of ingesting probiotics is to restore the population equilibrium in an ecosystem, like the reintroduction of wolves into Yellowstone Park. (A fecal transplant, for that matter, is also a probiotic. As the OpenBiome stool bank says, "rePOOPulate!"*)

Élie "death begins in the colon" Metchnikoff, a 19th century Russian zool-

---

*RePOOPulate is actually a synthetic stool composed of 33 bacterial strains developed by two Canadian scientists for fecal transplants.

ogist and Nobel Laureate, introduced the idea that gut flora could be modified by ingesting certain microbes. He observed populations of Bulgarians that lived long lives and determined the secret was their yogurt, laden with *Lactobacillus delbrueckii* subspecies *bulgaricus*. He drank sour milk every day and lived to be 71, a good deal older than average at the time, though unimpressive today (my Zumba teacher at the Y is over 70). It turns out *L. bulgaricus* doesn't survive the trip through the stomach, and a lot of other probiotics probably don't either. But that doesn't mean probiotics should be written off.

Probiotics are being studied for their potential to relieve inflammatory bowel disorder (IBD), irritable bowel syndrome, eczema, high cholesterol and blood pressure, bacterial vaginosis, and antibiotic-associated diarrhea, among other conditions, and they are in clinical use to varying degrees. There are claims the introduction of these bacteria can improve everything from cholesterol levels to autism—hence the new moniker, "psychobiotics"—and from infected burns to kidney stones. Probiotics in the *Lactobacillus* genus may help with weight reduction and someday may be used to treat obesity. But it's very hard to determine how effective they truly are or how much and what species a particular person needs.

Today, probiotics are commonly prescribed to patients with IBD, antibiotic-associated diarrhea, and infectious childhood diarrhea. They are also prescribed in conjunction with antibiotics, to treat urinary tract infections (there is a microbiome of the urinary tract) and bacterial vaginosis. Most patients in hospitals get probiotic supplements even though it is not really established that they work. But they are thought to do no harm unless a patient is given a really humongous dose—too much could cause bloating or give you diarrhea. Probiotics may do us a lot of good, or maybe they don't matter much. The science is still being sorted out.

Nonetheless, the probiotic market, which includes everything from yogurt drinks to supplements to animal feed, was over $35 billion in 2015, and it is expected to go up. "You can't underestimate the placebo effect," said Dr. Ari Grinspan, a pioneer of fecal transplants at Mount Sinai Hospital in New York City. People are buying the stuff without knowing if it works or whether or not they would benefit from a population increase of certain species if it did; nor are they analyzing the effects additives like sugar may have on their microbiota. Market analysts figure that as scientific studies determine the actual efficacy of a given probiotic, they'll do even more business.

Fermented foods like yogurt and kefir, sauerkraut and kimchi are characterized by the sourness of lactic acid, the by-product of lactobacillus fermentation. (Kombucha is sometimes advertised as a probiotic. It's a tea fermented by a culture of bacteria and fungi. Its benefits are even more specious than lacto-fermented products.) People who can't break down lactose can often manage yogurt or kefir because the lactose sugar in these products is partially broken down by bacteria. Foods that have undergone lactic fermentation have the dual benefits of preservation due to increased acidity, which wards off microbes that would otherwise spoil them, and the health advantages of fermenting cultures, which increase the nutritional value of the food (sauerkraut, for example, has more vitamin C than cabbage) and control intestinal infections. Since the bacteria in these foods are already adapted to an acidic environment, some may survive your stomach and could potentially set up shop in your intestine.

But the marketing of fermented foods, particularly yogurt drinks, is prone to hype. Commercial strains are not necessarily the same as research strains, or incompatible strains may be mixed together, or they could be simply crummy wholesale strains. Environmental conditions like warm temperatures can alter the community composition in your yogurt cup. Pasteurization kills many microbial species, and canning kills them all. In "probiotic" commercial yogurts, it would take some kitchen-table microbiology to determine the actual number of living cells, and even if you did, it's still hard to know how much of a given bacteria a person needs, as age, size, and condition all matter. Since there are no FDA standards, labeling is voluntary, and companies don't have to prove their products are effective in order to sell them. "The marketing of probiotics," wrote Dr. Martin J. Blaser in *Missing Microbes*, "is a kind of freedom of speech."

The best way to enjoy the benefits of fermented foods is to do it yourself, which is easy because, actually, all you have to do is allow the bacteria to do their thing. My friend Sean told me once he started eating homemade kimchi and goat's milk yogurt in the morning, "I am making such beautiful poops I have to look at them for a while before I flush them down."

A few weeks before graduation, I tried the Microbiome Diet, the "Scientifically Proven Way to Restore Your Gut Health and Achieve Permanent Weight Loss." The diet claims that balancing your microbiome "is the key to eliminating symptoms that you might never even have connected to your weight and the way you eat, such as fatigue, anxiety, depression, brain fog, headaches, acne, eczema, congestion, frequent colds and infections, joint pain, and muscle pain"—and existential strife (just kidding).

It starts with 2 weeks of vegetables, particularly those in the lily family, which bifidobacteria and lactobacilli prefer: leeks, asparagus, garlic, onions (my Italian grandmother used to say that onions calm the stomach), Jerusalem artichokes, jicama, radishes, carrots, and tomatoes. You can eat fruits, meat, poultry, and fish, but no dairy, grains, eggs, sugar, or alcohol, and minimal caffeine—kind of what lots of people do when we wing a diet. But add to this an array of supplements to the tune of $250. I got the Barrier Boost dietary supplement (which is mainly L-glutamine, an amino acid); butyrate pills (a short-chain fatty acid that nourishes the colon wall); a blend of plant enzymes; artichoke leaf and gentian root extract, both traditional GI treatments; the Microbiome Boost (includes tribulus extract—which you shouldn't take if you are pregnant or breastfeeding—and berberine, which actually might kill bacteria); Microbiome Balance pills, which consist of 100 billion hopefully live lactobacillus and bifidobacterium cells; and a prebiotic supplement made of powdered polysaccharides derived from the larch tree.

Prebiotics are bacteria food—specifically, the sugars in fiber that bacteria ferment for energy. Breast milk is our first prebiotic. Prebiotics are like the dead fallen leaves that feed soil microbes carbon and other nutrients. You could say prebiotics feed *our* inner soil. A prebiotic is the fiber in the apple that makes it past your small intestine all the way down to your colon, where the bulk of your gut microbes make a living. The fiber in an apple is pectin; in a carrot, it's amylose. These are the components of starch, hence the handle, resistant starch. There are other fibers, like the cellulose in celery and other plants, and chitin, which is what the cell walls of shrimp shells and mushrooms are made of. If you can't digest it with your own enzymes and the bacteria in your large intestine can, it's a prebiotic. Plants are our main source of prebiotics, including whole grains, though there are prebiotic supplements on the market. It's easy to be misled into thinking a prebiotic or a resistant starch is somehow a special new product. It's not. An apple is a prebiotic.

The more fiber you eat (25 grams for women and 38 grams for men a day—that's five large apples a day for me), the more you feed the microbes in your colon, the more useful by-products like fatty acids they produce, the healthier your colon wall, the less inflammation you suffer. The more processed foods you eat, like white flour and sugar, the more calories are absorbed in your small intestine. You aren't feeding your gut microbes; instead, you're dumping all those sugars into your bloodstream, where they cause trouble. And the more fats you eat, like steak and ice cream, the more you feed those microbes that are associated with bile (which helps the stomach digest fats), which leads to inflammation and colitis in mice and possibly in humans.

A meat-dominated diet feeds bacterial putrefiers that produce nitrogen and sulfur-containing compounds that damage the cells lining the colon. Chronic exposure of colon cells to these compounds may explain why colon cancer comes on late in life—like with smoking cigarettes and lung cancer, it takes a while for disease to appear—and it happens mostly in the lower part of the colon where putrefaction occurs. Additionally, if you eat lots of fats, you secrete more bile to break them down, which means more bile ends up in the colon, where microbes convert it into secondary acids that are also toxic to the lining of the colon. Eating meat is not the problem. The problem is eating more meat than vegetables. "So long as the byproducts of the fiber fermenters prevail," wrote David R. Montgomery and Anne Biklé in *The Hidden Half of Nature*, "then the colon serves as a medicine chest rather than a toxic dump."

Within a few weeks on the Microbiome Diet, I had lost 3 or 4 pounds, was pooping nicely, and thought my eyes looked a little less bedeviled. I think this mainly was because I'd switched to a plant-based diet. Diets high in plant starches feed bacteria from the phylum Firmicutes. In mouse studies, Firmicutes dominated in lean mice and Bacteroidetes, another phylum, dominated in obese mice. When the gut microbiomes of the mice were switched, the weight did, too. The gut microbiota of the mice differed in their efficiency in harvesting calories and differed in the way that harvested energy was used and stored.

Studies of humans were similar. Obesity is associated with phylum-level changes in the microbiome, not just a change of species, though diversity is also implicated. The Firmicutes-to-Bacteroidetes ratio is not a reflection of obesity; it is a reflection of diet.*

Our gut microbiota may influence our food cravings. The authors of an article in the journal *BioEssays* described the human gut as a place where different bacterial species are "under selective pressure to manipulate the host behavior to increase their fitness at the expense of host fitness." They may prefer a type of food that enhances their own fitness or suppresses the fitness of a competing species, and their demand for certain foods translates into our cravings. For example, a study found that people who crave chocolate may house different microbial populations than people with identical diets who don't care for chocolate. Gut bacteria may alter taste receptors so that their preferred food tastes better to us. But in a diverse microbial environment, the *BioEssays* article authors suggest that microbes will more likely "expend resources on competing and cooperating," like cross-feeding, where one species lives off the by-product of another, rather than on manipulating their host. The takeaway here is we experience more benefits from a diversified gut microbiota, and that's achieved by diversifying your diet.

My experiment with the Microbiome Diet was going very well until I decided to look up some of the supplements. I discovered we make glutamine every day. It's nonessential, meaning we don't need to get it from food we eat unless we are very sick with a disease like cancer. I also found out butyrate supplements are absorbed before they reach the colon, usually in the small intestine, which means the benefits to the colon cells would be lost. I have to say, that made me lose faith in the supplement side of the diet. To compensate, I doubled up on the apples.

You can call it the Microbiome Diet, or the FODMAP Diet, or the Zone Diet, or various vegetarian diets (lacto, lacto-ovo, pesco), the South Beach Diet, the Raw Food Diet, the Mediterranean Diet, but, ultimately, they all are about the same thing. If you want to lose weight and stay slim, eat lots of fiber, limit

*In a rather famous case, a mother suffering from a *C. difficile* infection got a fecal transplant donated by her daughter, who was obese. Mom's infection cleared up, but she promptly gained 40 pounds.

fats, and don't eat refined grains and sugar. And you have to do that *forever*. And it's best to start young. "Here's a diet for you," said mycologist and plant pathologist James White. "Go to the garden, pull out a carrot. Wipe it off and eat it. Eat it right there. That's the microbiome diet nature wrote."

James White told me that the Western diet is a great sterility experiment that has led to a lack of microbial diversity. We've overdone it, he said. The pendulum has swung too far, a reaction to the microbial origin of the most terrifying human ailments: tuberculosis, typhoid, cholera, plague, bacterial pneumonia, Legionnaires' disease, anthrax. But these aren't the diseases I worry about living in New York City. (Although Legionnaires' disease, caused by the proliferation of *Legionella pneumophila* in water systems such as steam rooms and mist sprayers in grocery stores, is a devastating lung infection that, like salmonella outbreaks and food recalls, tends to capture headlines.) What I worry about is cancer and Alzheimer's disease, illnesses not typically associated with microbial infections.

Except, of course, they are. Or they may be. Bacteria tend to accumulate at disease sites. Certain strains of bacteria thrive in a low-oxygen environment like tumors and in the favorable conditions created by the host's suppressed immune system. *Fusobacterium*, for example, has been associated with colon and breast cancer, but scientists don't know if it is the presence or absence of certain bacteria that can confer risk or lead to the development of disease. *Lactobacillus* and *Streptococcus* are more prevalent within healthy breast tissue than in cancerous tissue, and both genera produce anticarcinogenic properties like antioxidants, a service similar to that performed by endophytic fungi in plants. Some microbes seem to promote cancer, others may protect against it, and some seem to do both: *H. pylori* has been indicated in the promotion of one kind of cancer (stomach) and protection against another (esophagus). "There is no free lunch," microbiologist Moselio Schaechter told me when I asked him about this dichotomy. "That is the case of every symbiont. Anything can become a pathogen."

The American Society for Microbiology blog, *Small Things Considered*, conjectured that cancer cells could be considered an alternative organism somewhat like bacteria, a species all their own that grows in colonies, that cooperates and competes, "living protozoan-like fossils, foreign to their hosts because of deep origins elsewhere," said the oncologist Mark D. Vincent from the University of Western Ontario. We think of cancer as a series of cellular mistakes. Thinking of cancer as a species, one that is "oddly microbial," offers new directions for research and therapy, like quorum-busting strategies that may break up tumors, which are "akin to biofilms."

There is a growing interest in recruiting bacteria to fight cancer by creating living drugs, like programming harmless strains of anaerobic bacteria to deliver toxic payloads to tumors, a place where they like to hang out anyway. Or directly injecting bacteria into tumors, which might work because it stimulates an immune response in the vicinity of the cancer, even engineering bacteria to self-destruct in cancer cells, like microbial suicide bombers.

The other disease we all seem to worry about developing is Alzheimer's. There are a few hypotheses for its cause, but scientists at Harvard Medical School have hypothesized that the plaque known to clog the brains of patients may be debris left over from an immune response to microbial infections. The blood-brain barrier gets leaky with old age, the leakiest part being the membrane that protects the hippocampus, the site of memory. If a bacterium (or other microbe) makes its way across the barrier, the immune system kicks in and traps the microbe in a "sticky cage of proteins . . . like a fly in a spider web," reported the *New York Times*. The remnants of those cages show up as the plaque in patients' brains. The plaque ultimately causes the death of nerve cells that are related to memory functions. Whether you develop Alzheimer's or not may depend on your ability to clear out the plaque, which has a genetic component.

What those microbial infections of the brain could be are unknown, though a small study found that Alzheimer's patients have higher levels of bacteria from the Actinobacteria family, specifically, *Propionibacterium acnes*, which causes acne. But that doesn't mean *P. acnes* leads to plaque buildup. Maybe the relevant infections are just garden-variety types, like bronchitis or sexually transmitted diseases like chlamydia.

But it's the infrequent diseases that are the spookiest, like leprosy, the original horror bacterial disease. Caused by the bacterium *Mycobacterium leprae*, this ancient disease causes severe deformity and skin lesions. It's been all but eradicated, but it remains culturally terrifying. When I visited a friend one spring in the Missouri countryside, I found numerous dead nine-banded armadillos, which are curiously beautiful, like field seashells. I had collected quite a few of the sun-bleached armor plates when my host pointed out that armadillos are reservoirs for *M. leprae*. Even though I knew that 95 percent of people are genetically unsusceptible to leprosy, that it's highly treatable, and you won't get it just by touching an armadillo shell, I still took about 10 showers.

But far out-horrifying leprosy are flesh-eating bacteria. That may be because this gruesome infection seems to infect in the most random, mundane ways: a small cut on a boy's finger from taking off his football helmet; a boater wading in a river without realizing he had a scratch on his leg; a woman who cut her hand preparing a piece of tilapia.

"Flesh-eating bacteria get a lot of press because it's reasonably spectacular," said the mycologist Tom Volk, who for years battled *Staphylococcus aureus*. It wasn't looking for a human host, it was just Tom's very bad luck the bacteria traveled through his bloodstream and escaped into the tissue of his foot. "I'm underplaying it. It's actually pretty devastating." Flesh-eating bacteria consume the fascia, the sheath of fibrous tissue that wraps muscles, and their toxic by-products lead to a condition called necrotizing fasciitis, or tissue death, something like what happens when you get bitten by a brown recluse spider, only worse. The treatment includes massive doses of antibiotics and surgery, which involves scraping the dead tissue away, or, if necessary, amputation to stop the bacteria from spreading. Tom's infection happened in 2005. It's not progressing, but it still hasn't healed. He sent me a few pictures from the worst stage of the infection. I looked at them briefly, once.

Despite the headlines, infections of flesh-eating bacteria are rare. One out of 300,000 people a year become infected, and infection is more likely to occur in people who have weakened immune systems. There are several different species in several different phyla that cause the disease. What they have in common is they eat fascia and live at body temperature. A 2002 paper published in the *New England Journal of Medicine* described how German doctors were forced to amputate the thumb of a diabetic man who licked a small wound he got from falling off his bike and developed necrotizing fasciitis. It turns out two kinds of bacteria found in the mouth, *Eikenella corrodens* and *Streptococcus anginosus*, were responsible for the infection.

We all carry infections of some sort or other. If you have had chickenpox, you carry latent virus in your nerve cells that can reoccur as shingles, a painful skin rash. Tuberculosis may be latent in a carrier. There's a 10 to 20% chance an American hosts the bizarre parasite *Toxoplasma gondii* (the percentage is higher in other countries). We get it from cats, and it causes brain cysts that lead to subtle but bizarre behavioral changes.

There may be more to the cat lady phenomenon than just senility. *T. gondii* is a single-celled soil protist that needs a cat to complete its life cycle. To do this, it finds its way into a mouse, which is eaten by a cat, where the protist reproduces, and its progeny returns to the soil in cat feces. *T. gondii* increases the odds its mouse host will be eaten by exerting a kind of mind control. This is a technique other microbes use to move along their life cycles. For example, the fungus *Ophiocordyceps unilateralis* infects an ant and interferes with its brain-muscle coordination, becoming a virtual puppeteer. The fungus forces the ant to crawl into the foliage above the nest, whereupon it kills off the ant and produces a fruiting body out of its head that rains down spores on the ant colony. What *T. gondii* does is turn genes on and off in the mice it infects, making them less fearful of cats and so easier prey. We can get infected from cat feces, water, or rare meat. The disease, toxoplasmosis, can cause birth defects like intellectual disability and blindness, which is why pregnant women are not supposed to empty litter boxes. Getting the infection is something like getting the flu; you won't know you have contracted anything unusual, but the protist can affect your behavior in self-destructive ways. According to the evolutionary biologist Jaroslav Flegr, infected men become more aggressive and infected women become more trusting—risky behavior for each. Both genders are more prone to car accidents, and the disease is three times more likely to be found in someone suffering from schizophrenia.

Parasites can tinker with the mind, reported Kathleen McAuliffe in the *Atlantic* magazine. Maybe the flu virus inclines us to socialize, allowing the virus to spread to new hosts. Some people at the end stages of certain STDs experience intense cravings for sex. It begs the question, wrote McAuliffe, "who's running the show?"

But the litter box is the least of our problems. Here's what we should be fearful of. In September 2016, a female in her seventies died from an infection caused by a type of Enterobacteriaceae that was resistant to all available antibiotics. It is one of the 12 most dangerous bacteria, according to the World Health Organization, because nothing, or very little, can treat it.

We've enjoyed the benefits of antibiotics since Alexander Fleming discovered penicillin from a common bread mold in 1928.* It's likely that everyone in America has a family member whose life was saved by antibiotics, according to the authors of *The Hidden Half of Nature*. Antibiotics are small molecules that mess with functions in bacterial cells, like the ability to build a new cell wall or make proteins or divide, and by the 1960s there were hundreds. At the end of his Nobel lecture, Dr. Fleming noted that "there is the danger that the ignorant man may easily underdose himself and by exposing his microbes to non-lethal quantities of the drug make them resistant." And, indeed, by the end of the 1940s, streptomycin stopped working for some tuberculosis patients.

Only a handful of truly novel antibiotics have made it to market since 1980, and many of the largest drug companies have reduced or closed their antibiotic programs. They don't want to invest in figuring out how to find new antibiotics, not when patients get well after using them for 10 days (versus drugs like statins, which patients take every day for the rest of their lives, or cancer drugs, which are very costly). There's a fundamental marketplace problem with finding new antibiotics. We've been binging on antibiotics for decades, "carrying out a largely unsupervised global experiment in bacterial evolution," said the microbiologist Diarmaid Hughes of Uppsala University, which has led us to the brink of a postantibiotic era.

Antibiotic misuse has happened in the most banal ways: to fatten up livestock or administer a prophylactic against disease in overcrowded pens and allowing those antibiotics to seep into soil, to disperse in the air, to travel in irrigation water; to pacify parents by giving antibiotics to children with viral infections; by using antimicrobial soaps on floors in schools and then dumping the used solution down the drain. Every time we flush, we send a load of resistant bacteria into our sewage plants, where they comingle and share genes with other bacteria in an orgy of evolutionary selection. We've been pretty noncha-

---

*Not that it wasn't known in folk medicine. During the Civil War, injured soldiers slapped moldy bread onto their wounds. In 2017, a wad of mold from Dr. Fleming's lab fetched over $14,000 at auction.

lant about what life would be like without antibiotics, but here's a sobering projection. Antibiotic-resistant bacteria could kill 10 million people every year by 2050, according to the economist Jim O'Neill, costing $100 trillion to the world economy between now and then. Without effective antibiotics, surgeries like C-sections and appendectomies become more dangerous. According to the Centers for Disease Control and Prevention, at least 23,000 people die each year from antibiotic-resistant infections in the USA. It's higher elsewhere, and it is going to get higher everywhere. So, what to do?

Experts recommend we improve sanitation like sewage systems and keep air clean, as air pollution transmits antibiotic-resistant bacteria. We need to regulate nonessential antibiotic use in animal husbandry. Currently, curtailing prophylactic antibiotic use in dense growing conditions is voluntary. We need to reduce our use of antibiotic household products.* We need to create faster diagnostic tools so doctors don't have to give antibiotics for what may turn out to be a virus. One avenue of investigation is looking at the patient's immune system—we respond differently to bacterial infections than viruses. And we should do a better job of explaining to parents why their children don't need an antibiotic for every sinus infection (viruses cause yellow phlegm, too). Your mother doesn't need antibiotics when she gets her teeth cleaned just because she has a heart murmur, and your newborn doesn't need antibiotic drops in her eyes unless you have an STD.

We also need to develop alternative drugs, like phage therapy, which uses bacteria's natural enemies to control infection, and vaccines, which are more effective than antibiotics anyway, and consider new ways of attacking bacteria, like their ability to quorum, or look to how other animals deal with dangerous microbes, like the antibiotics that leaf-cutter ants grow in their nests or the antimicrobial amino acids in the milk of Tasmanian devils. (If you've ever met the utterly unpleasant Tasmanian devil, it seems plausible their milk kills.) And we need to pay for new drugs, which means supporting legislators who are willing to think a generation ahead and use some of our taxes to finance the research and development. A new generation of antibiotics is in the public

---

*There's some good news. The use of antibiotics to fatten livestock is on the wane, and the FDA banned the sale of soaps containing antibacterial chemicals in 2016; it turns out antibacterial soaps aren't more effective than regular soap and water anyway.

good, said Dr. Blaser, "like road building." It's an investment in our children
and their children; it's insurance against an era of grief.

Columbia University has two commencement events: Class Day, which is particular to each of Columbia's schools, and University Commencement, when
students are formally conferred their degrees. Columbia has hosted many
famous Class Day speakers, but Barnard probably tops the list by hosting a sitting president, Barack Obama (whose alma mater is Columbia College). This
reignited a long-simmering sibling rivalry. Male Columbia College students
accuse Barnard students of getting a Columbia University degree through the
backdoor because Barnard is less competitive, and Barnard students accuse
Columbia College boys of being whiners and lousy in bed. One Barnard student
blogger pertly recommended Columbia settle the score by asking a Barnard
alumnus to speak, maybe Martha Stewart? Which really pushed some over the
edge into the unprintable. But there was a point of cooperation. Students from
both schools happily ganged up on a Harvard troll who ventured a comment
along the helpful lines of you are both losers, so "suck my crimson dick."

I attended University Commencement on a sunny day in mid-May. I joined
a stream of people, parents mainly, dressed up in summer suits and skirts, in
saris, kimonos, and dashiki, as we were pressed by blue-shirted security officers
with megaphones—one entertained himself by saying "Chicago Cubs fans
only"—who herded the crowd along 114th Street between metal barriers,
toward the side entrances to the campus. It was like going through a snaking
security line at the airport, only everyone was carrying flowers. We grabbed
water bottles, courtesy of the alumni committee, with labels that read "330,000
Columbia alumni worldwide, one powerful network," and a program, and a
copy of the student newspaper, the *Columbia Daily Spectator*, and we were
reminded that all exits were final. Once our ticket had been scanned, we
couldn't leave and reenter the ceremony. It occurred to me that I actually had
exited once, 25 years ago, and here I was. Reentered.

Around 30,000 of us found chairs on the lawns, facing the huge cement
dome of Low Library (the largest, the megatron screens told us, in the USA),

and while the "Pomp and Circumstance" march played over the loudspeakers, the students filled their seats, a carpet of pale blue robes and balloons and waving flags laid before the imposing library steps. Assembling all the students and guests who were to receive honorary degrees was an hour-and-a-half proposition, and in the bright sun, parents began to wilt—the handy among them made paper pirate hats out of the student newspaper to offer some shade—and I wondered how the students were doing in their 100 percent recycled-plastic robes. The music changed for the academic procession. While the tangled Medieval harmonies of lute and harpsichord and horn played over the loudspeakers, the faculty, administrators, marshal, chaplain, and other dignitaries marched to their seats, all in their regalia robes and mortarboards (the black, or in the case of Columbia blue, tasseled flat-topped cap) and pouchy Tudor bonnets.

The last dignitary carried the school mace. Many university commencements parade their school mace, a type of medieval weapon used to bash people's heads in. Today, it is a symbol of the school president's authority to confer degrees. Columbia's mace is a wooden club with a carved silver head, topped with a crown. The UK's House of Commons has a mace, too. Parliament cannot lawfully meet without the mace present, which represents the monarch's authority and has occasionally been seized and wielded by passionate members of Parliament. The USA has had a mace in the House of Representatives since 1842. It's an ebony rod with a silver globe and winged eagle on top. When the house is in session, the mace stands on a green marble pedestal to the Speaker's right. When the house is in committee, it's moved near the Sergeant of Arms' desk, and if a member becomes unruly, the Sergeant of Arms may lift the mace and show it to the offender to get them to shut up. This happened in 1994 to Representative Maxine Waters of California, who was arguing an issue of women's rights. She left the podium before the Sergeant of Arms could "present the mace," that is, silence her by threatening to symbolically bash her head in.

Medals and honorary degrees were handed out, and then each dean came to the podium and pitched their graduate class, a symbolic plea to the president to confer degrees. The deans carried on about how prepared their students are, how smart, how talented, how persistent they are, and how much they've learned, including, according to one dean, talents like "the Jedi mind trick where they tell their families, 'I'm almost done with my thesis,' and the families repeat, 'She's almost done with her thesis.'" It was all so joyous and grandiose. When I bailed out on my own graduation, I forfeited my small part in this theatre, and as I sat in the audience, I tried to remember what I was so sour about. I suppose I was just ready to let living be my education, and then I forgot what the academic experience felt like.

Going back to school this time around involved the disassembly of my ego, never easy. But ironically, or maybe fortuitously, one has to let go of one's ego in order to learn anything about biology. Ego, and all the hubris that comes with it, blinds us to the extent of nature's complexity. It's ego that allows us to think we see the whole picture; when in truth, there is a lot more to life than meets the human eye. For me, that lesson was the most valuable one of all.

So it didn't matter that I was watching graduation for the first time from the bleachers. It was wonderful being there, sitting behind a man who kept saying to his family, "He did it! He did it!" and mouthed, "I love you" as he kissed his wife. The ceremonies closed with the chaplain wishing the students "safe travels" and "Godspeed," and then they all let loose their balloons and threw a thousand caps in the air and the loudspeakers started to play Frank Sinatra's "New York, New York," and for some reason that I think had to do with the feeling that I was a part of something bigger than myself, the community of learners, perhaps, I started to cry.

After the ceremony, the Columbia building department workers got to it. On my way to the exit, I had to negotiate a maze of barriers and bleachers that were continually changing as they hustled to disassemble the stage, not to mention the obstacle course of parent-graduate photo ops, selfie-stick groups, and lost people wandering around. I had a déjà vu moment, and I paused to take in the hugging, proud, swirling crowd before heading into the dark anonymity of the subway.

A few months after graduation, I took a train to Boston to attend the American Society for Microbiology Microbe 2016 conference. As I streamed into the conference center with 10,000 microbiologists, a catalog of lectures in hand, pencils sharp, it occurred to me that maybe the reason why I didn't go to my graduation was because I never saw myself as being done learning.

CHAPTER

12

The Earth Microbiome

I stayed with friends in Cambridge during the American Society for Microbiology's (ASM) 3-day conference in Boston. Susan Goldhor is a biologist
who I know through our mutual interest in mushrooms. She is the head
of the Boston Mycological Club, the oldest such club in the USA. The club hosts
mushroom-hunting walks and talks like Mushroom Poisoning: Myth, History
and Medicine and social events like the annual Duff Sale, a kind of Black Friday
sale for fresh and dried mushrooms.

Susan sent me directions that were achingly accurate, just short of how
many steps to take from the top of the subway escalator to the main gates of the
Harvard campus, though on paper they seemed totally obscure. And yet, after
I "cut right into the Bio Quad," I did see the "two rhinos and other zoological
stuff" that identified the biology building. I still got lost and had to stop a student. When he turned his head to point me in the right direction, I noticed he
had hickies on his neck.

On the other side of campus, on the first narrow residential street, I found
Susan and her husband Aron Bernstein's '70s-era wooden row house, embody-
ing the decade's wealth of contradictions in its combination of earthy homey-
ness and groovy colors. Susan and Aron are intellectuals with '70s radical
streaks. His passion is nuclear arms control—Aron is a physicist at MIT—and
hers is the manly pursuit of seafood waste management. They are petite and
opinionated and cozy, the kind of people whose end of the dinner table you
want to sit at. I stayed in a basement room with a rickety bookshelf that held a

diverse collection, from Studs Terkel to soft porn, and my very own entrance, from which I slipped out each morning, retracing Susan's instructions backward, and made my way to the subway and the conference center.

The ASM Microbe conference is where you find out what's new in microbiology. I picked up a lot of biology in my year at Columbia, but if I wanted to concentrate on microbiology, I would have had to first study my nemesis, chemistry. I was momentarily tempted by the University of Florida's microbiology master's degree online because they really seemed to want to make it easy for me to sign up. But I've been the online route before and had problems, and anyway, attending the conference was much cheaper, $264 versus $16,050.

The conference was actually a merger, bringing together two groups, the Interscience Conference on Antimicrobial Agents and Chemotherapy, which is focused on the clinical side of things, and the ASM's general research community. Eleven thousand people were registered, more than the population of Mykonos.

The first event, late Thursday afternoon, was a talk by Bill Gates, who had just committed $100 million to the National Microbiome Initiative, which studies the microbes in humans, crops, soils, and oceans. The keynote was supposed to happen on Saturday night, but "when Bill Gates says yes," said the conference's director of meetings, "you take him on the days he says *he's* going to speak." I followed streams of people disembarking from subways and buses and headed through the great glass doors of the 516,000-square-foot building with its awning that looks like the bill of a giant steel Red Sox cap. Huge vertical banners decorated with dancing colored rings like a denatured Olympic symbol read "ASM Microbe."

Everything was, unsurprisingly, well organized. There were orderly lines to pick up our badges and very disappointing swag bags. Inside were just glossy pamphlets for businesses like Plas-Labs, the small animal catalog, the cover a collage of a double helix with a mouse, a rabbit, a chicken, a weasel, and a monkey perched on its rungs, and Tetracore, which makes BioThreat Alert test strips (among those available: anthrax, plague, ricin, and botulinum toxin). We also received the ASM Exhibit and Poster Hall Activity Guide, an index of the thousands of scientific posters presented at the conference, like the posters you did in school for the science fair, only really advanced. There were alleys of these posters, the "consolation prizes for people whose research didn't warrant a presentation," said one disgruntled PhD, pinned to rows of temporary walls in the main exhibit hall. The posters surrounded 250 exhibit booths that filled the central piazza of the hall, promoting everything from lab research equipment like the Anoxomat anaerobic system, "the precise automated solution for bacterial culturing, in a jar, in less than 3 minutes," to Oxford University Press, the preferred publishers of the super smart.

Gates's talk was held in a giant auditorium that we accessed in security relays. I followed the crowd: show your badge, walk along a two-story-high corridor for a quarter mile in one direction, show your badge to security again, take the escalators down, show your badge again, walk left down another long corridor in a river of people but against a tide of people coming in the opposite direction, until you reach a security barrier where the officer sends you back to the escalators to turn right instead. The scale was so large it didn't feel like individual navigation was even an option. I was a fish in a school. Finally, get in line to check your backpack, a requirement of the Gates security team. The line snaked down the corridor, a hundred scientists standing patiently, checking their ASM apps. I folded my backpack to make it look like an ugly purse and ditched the line.

The auditorium was like a cross between a classroom and an arena. The crowd found seats facing a stage while music blared in synchrony with megatron LED screens that flashed images of giant bacteria, big as hogs, and spiny viruses, the graphics splashy as the jumbotrons at a ball game. The audience was varied: all races, all ages, from college students to great-grandpas, as many women as men, and a dozen languages spoken. It was like walking the streets in diverse Queens, only everyone knew what *metabolism* meant.

When the ASM president Lynn Enquist, a molecular biologist at Princeton, walked across the stage, we applauded, but the murmur of voices in the vast room never totally abated. It was too large and too public a space for anyone to capture the attention of all. "Wow," said Dr. Enquist in the mild tones of a kindergarten teacher at a parent meeting, "this is exciting."

Over 6,000 of us were in that room, and we were all sharing microbes. As a species, we have a fundamental microbiome (with differing ratios of particular phyla), but each person's is also individually unique, and the diversity of individual skin microbiomes is quite high. We each travel through life in an aura of microscopic debris that includes viable, dormant, and dead microbes and microbial parts, and we are shedding this debris all the time, mainly from our skin and hair. In a small study published in *PeerJ*, a free online journal, human microbial "clouds" were found to be different from background airborne communities of microbes, and each human cloud was unique. Women's clouds were different from men's—it includes common vaginal bacteria—and different from each other, and the abundance of microbes in the clouds varied, a reflection of the amounts of bacteria shed by a specific person. It's long been shown that men shed more bacteria into their surrounding environment than women, but that may be because of differences in hygiene, skin health, or perspiration. The source of some of that variability is pretty obvious, though. It doesn't take a measuring device more complex than your nose to determine the microbial cloud of a toddler fresh from the bath is very different from the microbial cloud of a homeless man napping over a steaming manhole. Our microbial clouds are short-term pools of microbes in a given space. It's ephemeral. Because when we leave the space, our cloud settles to the floor, to be stirred up with other abandoned clouds when someone else walks through what was part of us.

When humans occupy an enclosed space, we contribute airborne bacteria. In a study of a university classroom, twice as many bacteria were present when people were in the room. Every person who enters adds to the microbiome of that space and depending on how much time and what degree of contact they have, two or even more people's airborne microbial clouds may homogenize. As James White suggested, shared microbes are one definition of family.

In every household, there is a different microbial composition, unique to that family, and those microbes are mainly coming from the people who live there. An only child is more likely to have hay fever than one with three or four siblings, and the reason is thought to be lack of exposure to the additional microbes other siblings might carry.

When you move into a house, your family's microbiota quickly colonizes the lingering microbiota of the family that lived there before you—in as little as a day. It doesn't take long for a house to become your home. Sharing microbes can be benign, but it can also be a vector for transient microbes like a virus, where people in the household share a cold, or a shared household microbiome can reflect the presence of microbes associated with chronic ill health, like obesity.*

What is the first thing you do in a new house? I open the windows. Open windows allow new species of microbes in, which diversifies the airborne bacterial communities inside and mixes the human microbes with outdoor microbes. Outdoor microbes reflect the land and how it's used, whether a city or a forest or a cornfield, connecting the occupants of the family with the geography of where they live.

What is the next thing you do?

I clean. There are different bacterial communities in different parts of your "houseome," as the geneticist Rob DeSalle calls it, like the bathroom or the kitchen. That's not unlike how geography, such as bodies of water or altitude, affects the structure of plant and animal communities. Within a particular part of our houseome, like the kitchen, different surfaces harbor different kinds of bacteria, probably in association with how those places are used, similarly to how the bank structure of a stream is associated with certain species of fish. There is an evolutionary relationship between an environment and the organisms that live in it, even if all we are talking about is the environment of a kitchen sponge.

Food magazines remind us to use one cutting board for meat and another for vegetables, to microwave our sponges, but a study in 2012 found that human skin is the primary source of bacteria across kitchen surfaces. My friend Alex is scared to death of raw chicken juice contaminating her countertops, but the study also found common food-borne pathogens, while all over, were low in numbers. When you wipe a counter, what you are really wiping up is the microbe's food, the bits of grease and residue of juices they eat, and that's usually adequate to thwart a problematic increase in pathogenic microbes. Antibacterial soaps and sprays containing the antibacterial compound triclosan are not more effective than soap and water. They make things worse. They add to the antibiotic load in our environment—already triclosan is present in most

---

*One study found that it's possible to detect if someone has been in a room by the presence of his unique microbiota on objects like door handles and computer mice for up to 2 weeks—a fact with forensic value.

American freshwater streams and rivers—and encourage the evolution of antibiotic-resistant bacteria.* If you want to kill bacteria, use rubbing alcohol, which does the job and then dissipates.

All the occupants of a home play a role in determining the types of bacterial communities found indoors, and that includes your pets. From the microbiology alone, researchers could tell with over 83 percent accuracy whether animals lived in a home. Dogs have diverse microbiomes from their hair and their habits, as evidenced by a recent walk on the beach with a friend's hound. She rolled in the carcass of a dead seagull and ate the limp gnat-covered crabs that washed ashore and then spent the afternoon napping on my friend's couch. The constant shedding of your dog's skin microbiome may cause allergies or the opposite. Having a dog in the household is associated with a decrease in allergies in children, as is childhood exposure to farm animals. Pregnant women may be surprised to learn households with dogs have even been found to positively affect the maturation of their baby's immune system while he's still in the womb. (The dark side of living in proximity to farm animals is the ability of some microbes to jump species: smallpox, measles, mumps, diphtheria, whooping cough, and scarlet fever were originally animal pathogens. In the 21st century, diseases like swine flu and avian influenza are thought to have arisen from proximity to farm animals.)

But allergies to microbes like mold spores are a real problem for as many as 50 million people in the USA. They can trigger typical allergic reactions like sneezing and wheezing and an itchy throat in otherwise healthy folks and asthma symptoms in susceptible patients. No one knows how many American homes harbor molds, but according to the International Center for Toxicology and Medicine, the best current estimate is about 70 percent. The EPA points out there is no way to eliminate all molds and mold spores in an indoor environment, any more than you can eliminate bacteria. In 2001, a Texan won $32 million against her insurance company—later reduced to $4 million—for mold damage to her 12,000-square-foot home, a replica of Tara, the mansion from *Gone with the Wind*, in Dripping Springs, Texas. This led to a "mold rush."

---

*Triclosan is in a lot of products, from CVS antibacterial soap to Ajax antibacterial dish liquid, from toothpaste to lip gloss and hair detangler, from deodorant to shaving gel, kitchenware, Merrell shoes, Playskool children's toys, and a host of others.

Over the next few years, mold-related litigation tripled in some states. As one national legal group advertised, "If mold or fungus problems are in your home or condo, you need to make sure that you don't allow these issues to get worse. Contact a lawyer today." Despite the lawsuits, there are no US regulatory standards for indoor levels of mold spores. There is not even a generally accepted definition of dampness, or what constitutes a dampness problem. Fungi in homes tend to come from the outside, so "if you want to change the types of fungi you are exposed to . . . then it is best to move to a different home (preferably one far away)," wrote the authors of the paper "The Ecology of Microscopic Life in Household Dust." "If you want to change your bacterial exposures, then you just have to change who you live with," they added.

Subways aren't the petri dish of dangerous germs that most of us imagine. They actually have unremarkable diversity. Subway microbiomes, in New York City anyway, are mostly a combination of the same fungi that's floating around outside and human skin bacteria. And like kitchens, surfaces with specific uses house the expected bacteria. For example, researchers found turnstiles abundant with *Acinetobacter*, as well as *Enterococcus* and some *Streptococcus*. All three are common in feces. All three are also common on your butt-wiping hand (as *New York* magazine put it, "the whole world is covered in poop"). There is an element of truth that grabbing a subway pole is like shaking hands with a million people, it's just those million people aren't that different from you.

In cities, the airborne microbiota may homogenize as well. In a study of nine offices in three American cities, researchers found that offices have city-specific bacterial communities, "such that we can accurately predict which city an office microbiome sample is derived from." Even though a city is not an enclosed space, and the microbiome sample of Detroit may be different from one day to the next, there may be regional microbial populations that reflect geography and weather. For example, a study of airborne microbial diversity patterns found coastal regions in the United States share similar airborne microbial communities.

What, then, is the extent of the human microbiome? The individual's microbiota, which can homogenize with a lover's? A couple's microbiota, which can homogenize with their children's? Even with their pet's microbiota? Can our understanding of the human microbiome be extended to our cities? Our regions? In an essay from 1958, the sociologist Donald MacRae wrote, "The last word of biology is the first word in sociology." So it is. Because how connected we are by microbes is an indication of how connected we are to each other.

Bill Gates, rumpled and lanky, sat in a chair facing the well-groomed and tele-visual Richard Besser, the health and medical editor for ABC News at the time, formerly of the CDC. Gates's audacity in the face of global disease is dazzling. The Bill & Melinda Gates Foundation works for global health equity, that a child in a poor country should have no more of a chance of dying than a child in a rich country. At the prodding of his interviewer, Gates explained that the highest financial return in global health is vaccination—no need to get into the vaccines-cause-autism conspiracy with this crowd—because if you succeed in eradication, you save on all the costs of caring for those who have the disease for a lifetime. But as we approach the total eradication of a disease, it becomes more expensive. "The last 1 percent is very hard," said Gates. "Per case, at the end you spend an immense amount to get all the years of zero [cost] in the future." For example, two countries remain before polio is eradicated, Pakistan and Afghanistan. Those countries have few cases, but it costs twice as much to immunize those last subjects because the Taliban targets polio workers, and the children that aren't vaccinated become a potential pool for further spread of the disease. But he also explained that if there was one single most important prob-lem that he'd like to address, it is nutrition, and nutrition, including malnutri-tion and obesity, is intimately tied to the human microbiome. "The opportunity to understand the microbiome," he said, "to have deep diagnosis, is critical to achieving our goals."

A murmur 6,000-people strong rippled through the auditorium, and I imagined 12,000 fingers pulling their microscopes into focus. Mr. Gates knew his audience. Microbiomes—those shifting yet coordinated assemblies of microbes that underlay the functioning of all organisms on Earth—that was their thing.

After the keynote talk, the crowd exited and I exited, too, as if being disgorged from Fenway Park, heading to the reception. We walked outside the conference center, and it seemed like instead of walking one-quarter of the distance around it in one direction, we walked three-quarters of the distance in another, and more than once the person that I was walking beside for a while asked if I had any idea where we were going.

We finally poured through the classical archway entrance of the Seaport World Trade Center and its piazza. Hundreds of people were lined up at a variety of mass-catered food tables where the carrots and cauliflower and celery all were cut into lunchbox portions, the cheese cubed, the meats curled, the crackers spread like decks of cards. There were wine stations, too. I paid for a glass of $4 chardonnay with a $5 bill, and the bartender kept the change, avoiding my eyes. The crowd was huge and fluid, and as I wandered around with my plastic cup, I felt socially adrift, as if I'd missed my train and had hours to kill before the next. I practically pounced on a microbiologist from North India who was resting on the cement lip of a potted tree. Arunima studied oral bacteria, and I asked her what she hoped to get from the conference. She explained she was there to find someone to discuss problems she'd encountered with her research. "It's like that," she said, and not much more.

Throughout the conference center, there were stations with a coffee machine and a few couches called Peer-to-Peer Exchange zones, and I imagine Arunima parked herself in the zone Host-Microbe Biology. There were speaker-connection zones, too, where folks with briefcases sat side by side, typing on their computers, so if the speakers were connecting, I guess it was online. There were no bars where you could start up a conversation under more casual circumstances, although there was a row of chairs with foot massagers where I chatted with a beautiful red-head named Erin Symonds, an environmental microbiologist at the University of South Florida who said she had a poster in the exhibit hall but failed to mention she was to receive an award for research excellence and potential at the end of the conference. When I recognized her on stage, I jumped up and cheered loudly. She was the only person I knew.

That's not totally true. Microbiologist Moselio Schaechter was there. He was a plural presence, traveling in a constant cloud of aging colleagues and eager students. Everyone wanted a piece of Elio. I did wrangle a brief lunch where he told me that the public interest in our microbiomes had validated microbiology and placed the discipline where it belonged, not as the science of tiny living things but as "the science of life."

I didn't suffer too much from a lack of social interaction, however. I had a packed agenda. I listened to talks like Tracking the Archaeal Origins of Eukaryotes; The Human Gut Resistome; The "Normal" Human Microbiome; Gut Bacteria in Stink Bugs Confer Pesticide Resistance; Friends with Benefits: Gut Microbes Allow Herbivores to Ingest Toxic Plants; The Root Microbiome, during which the speaker, Jos Raaijmakers, said that what scientists are learning about microbiology is more like stamp collecting—it's in bits and pieces; Tracking Resistance between Humans, Vegetables, Fruits, and the Environment; More Trouble with Ticks; and Feed Me, Feed My Microbiome, in which the presenter, David Mills, a microbial ecologist, pointed at the screen and said a little wistfully, "This is 15 years of work in my lab summarized on one slide." But there were a hundred more presentations that I missed.

I noticed quite a few talks had jazzy titles, just like we'd been taught to do in lab. Friends, Foes, and Frenemies; Hitchhiking across the Eukaryotic Host Cell; Let's Start at the Very Beginning (a nod to *The Sound of Music*); and Meet Me at the Membrane (a nod to another musical, *Meet Me in St. Louis*). I bolted from one presentation to the next, running sometimes if the distances were great. The conference center was icy cold, so cold that my fingertips were constantly white, as if dipped in milk, a climate modified, said my host Susan when I mentioned it, for the suit-and-tie. There were cops everywhere, which was surprising. This is hardly the crowd to commit a crime, although collectively I think they could have prescribed more OxyContin than was currently on the streets of Boston.

I took frequent breaks, either intentionally or because I often got lost, to wander the exhibit hall and admire the displays of microscopes, the objectives softly clicking into position, the desktop PCR instruments, and the laboratory autoclaves for sterilization. I followed one ambling scientist who hit a number of exhibit booths, pocketing the pens they gave away, not the pamphlets you were morally required to take as well. He was oblivious to the salespeople's scowls, but I smiled at them because I felt bad, and then they'd try to get me interested in trying out their anaerobic chamber, where you stick your hands into rubbery sleeves to work in an airless box just like in the movies. But I begged away, embarrassed, not wanting to waste anyone's time. I lingered at the LabRatGifts table, as did a surprisingly large number of people. They were scarfing up the bumper stickers that said Honk if U Love Microbiology and Trust Me, I'm a Microbiologist; the ties depicting the table of elements and floating double helixes, and plushy toys of bacteria like *E. coli* and viruses like the flu, to take back to the kids at home.

I also stumbled upon the Agar Art show, which was tucked in a long hallway on an upper floor. Agar is the gooey stuff in petri dishes into which you add nutrients that encourage bacteria to grow, and agar art is the rather strange practice of growing colored bacteria in patterns in petri dishes. Alexander Fleming, the discoverer of penicillin, loved to create pictures of ballerinas on his agar plates, and microbiologists have been noodling around with this weird artistic medium for years. In 2015, ASM launched its first Agar Art show. There were 83 entries, which included the Chicago skyline made from a microbial species isolated from human lips and a superhero made from *Streptomycetes,* which can produce antibiotics. The show I saw featured a remarkably accurate portrait of Louis Pasteur made from the bacteria that can break down tarballs, and my favorite, "Dearly Beloved," a purple depiction of the rune used by the late funk artist Prince when he was battling his record company, grown from *Serratia marcescens.** When asked about the challenges of artistic collaboration with living organisms, one microbe artist, whose "Garden of Gut Bacteria" depicted a landscape grown from, yes, human gut microbes, grumbled, "Never work with children and animals . . . or bacteria."

I also caught up with Dr. Sheldon Campbell, a professor at the Yale School of Medicine, aka the singing microbiologist. He led a slightly awkward but utterly hilarious sing-along. To the tune of "When the Saints Go Marching In," he sang, "If you look into your red cells this creature may be seen. But babesia

---

*In 1819, Italian peasants in Padua went berserk over what they thought was an outbreak of cursed polenta—it seemed to bleed. An Italian pharmacist named Bartolomeo Bizio investigated the polenta and concluded the red pigmentation was a by-product of a fungus. It was actually the pathogenic bacteria *S. marcescens,* which the US government tried to weaponize in the early 1950s (the heyday for kooky warfare ideas, like the *Botulinum* bacteria–tainted wet suit designed to infect Fidel Castro). The feds burst a few experimental balloons filled with *S. marcescens* over San Francisco Bay, causing a minor respiratory outbreak. Today, we know *S. marcescens* bacteria are one of the top 10 causes of hospital-acquired respiratory, neonatal, and surgical infections, and they are common in dirty contact lenses, where they can cause infection. They usually make their homes in guts, water, and soil.

can be handled as long as you've got your spleen. 'Cause when the ticks, come marching in, when the ticks come marching in . . . " Dr. Campbell uses the songs in lieu of a summary slide in his classes. He has a song about fungi and another about TB and parasites. As he explained on the podcast *MicrobeWorld Video*, "I'm not sure I can scientifically prove there is an improvement in school performance from the music, but every so often I have a 3rd or 4th year student who says because of your song I remember how to treat Lyme disease."

The conference offered updates on topics like the microbial origin of life and the microbes that link the carbon, nitrogen, and phosphate cycles; the complex ecologies that are the microbiomes of soil and plants; and the many surprising new connections scientists are finding between our gut microbiota and who we are and how we feel. While I often felt like I needed a translator (a typical poster title: "Oxidation of Cytochrome 583 Is Rate-Limiting When *Acidiplasma aeolicum* Respires Aerobically on Iron"—what the fuck does *that* mean?), I loved it. The ASM conference was like being in the midst of a giant historic mosaic in the making, a thousand tiny bits of information all being assembled into a tremendous mural of life. And true to that endeavor, the mood was serious and earnest and focused on the craft of moving toward a fuller picture of nature, one tile at a time. I found the conference a life-tumbling experience, but I think what was most unexpected was a talk by the microbial ecologist and plant pathologist Cindy Morris, about microbial meteorology. "Rain," she said, "is not only life-giving."

"It's alive."

In the late 1970s, a plant pathologist at Montana State University named David Sands thought he had eradicated a case of bacterial leaf blight with a seed treatment. The seed was planted, but a month later the field was covered with the bacteria. How did it get there?

It had rained.

That's when he got the idea for the bioprecipitation hypothesis. He and his research colleagues found that certain bacteria that live on plants are swept into the atmosphere, where they act as a nucleus for water molecules to organize around. A cloud is composed of ultralight droplets of liquid water that remain suspended in air. When these droplets arrange around a nucleus, they are capable of holding more water molecules, of being bigger. Soot and dust can nucleate rain drops as well as biological nuclei, but water droplets with a biologic nucleus like bacteria can freeze into ice crystals at a higher temperature than water droplets with nonbiologic nuclei, or no nuclei at all. Freezing allows the crystals to hold even more water molecules. When the ice crystal becomes heavier than the air column that is holding it up, it falls and melts on the way down as rain or refreezes to become snow. While the Soviets found microbes at the farthest reaches of Earth's atmosphere in the 1970s, the majority are in the troposphere, 4 to 12 miles above sea level, where most of the mass of the atmosphere is, including 99 percent of Earth's vapor. It's where weather happens.

And it's not just bacteria that nucleate rain. Every cubic meter of cloud contains tens of thousands of microbial cells: fungi, algae, and viruses, too. Rain forests produce millions of fungal spores that seed clouds. Pristine rainforests, said researchers at North Carolina State University, are "biogeochemical reactors" that produce fuel for rain clouds.

Depending on their size, airborne microbes can stay aloft for weeks. In a matter of days, air that brings rain and storms can also carry microorganisms across a continent or an ocean. Dust storms can shuttle unique microbial communities across the land, like a jumbo jet full of tourists. Some microbes don't survive high altitudes and get zapped by UV light or desiccated by lack of water, but others are well adapted to wind dispersal, like the spores of fungi. All species in the fungal genus *Fusarium* are ice nucleators. They are also some of the deadliest plant pathogens, like fungal rust, a lineage of which, Ug99, threatens global wheat production. According to the Food and Agriculture Organization of the United Nations, at least 81 of 101 economically important crop plant diseases are caused by species of *Fusarium*. But in clouds, bacteria are more abundant than fungi.

Various bacterial species nucleate rain, but the best described and most commonly found is *Pseudomonas syringae*. "Rain is part of *P. syringae's* life cycle," said Dr. Morris. It's a relatively weak plant pathogen (though not all strains are pathogenic) that is dispersed by rain splash on plant leaves. The bacteria can also nucleate ice on plant leaves, and at higher temperatures than normal frosts. Under cold circumstances, *P. syringae* can freeze the tissues of the plant, making it more susceptible to infection by other microbes. Freezing doesn't hurt the bacteria, because *P. syringae* contains antifreeze proteins. One way scientists have combated this problem is to produce colonies of genetically modified *P. syringae,* where its ice-nucleating protein has been snipped out, and then the modified bacteria are introduced in order to displace the natural *P. syringae* population.* When there is no rain, the modified bacteria can be swept up into the atmosphere, where they can nucleate raindrops, which helps the bacteria move to new hosts on the ground.

Dr. Sands imagined seeding rain clouds with *P. syringae.*[†] "We can plant seeds that get along with *P. syringae,*" he said, "and provide a rain nucleation source. Instead of seeing clouds pass over for 30 days, we can get rain. I think when a farmer grows a crop of food, he is also growing a crop of bacteria that will pay it forward downwind to the next guy to grow his crop."

Microbes affect weather even over the oceans. Microscopic algae produce an organic compound that includes sulfur (called dimethylsulfoniopropionate, or DMSP) that helps keep their cells intact. When the algae get stressed from sun, they make more DMSP. As they age, they convert the DMSP into a gas, dimethyl sulfide (DMS), and when they die, 50 million tons a year of this stuff rises into the atmosphere, where solar radiation converts it into a sulfate, a salt. The salt particles nucleate water vapor, and clouds form. The clouds reflect solar radiation and cool the surface water with shade, which de-stresses the algae. The production of DMSP is a response to heat that ultimately produces a cooling effect.

These elegant feedback loops between microbes and the atmosphere do contribute to the overall well-being of their biome. Microbes are one factor in climate regulation. How significant a global factor is unknown, but it may be a good idea to remember microbes have had, and continue to have, an impact on the atmosphere. They oxygenated it, and they are still producing oxygen, along with plants, and microbes are the only biological source of methane gas. (They

*Dead *P. syringae* are used as nuclei for snow-making machines. You can buy it by the bag. It's called Snomax.

[†]Cloud seeding has been attempted synthetically as well, with chemicals like silver iodide and substances like salt.

cycle foraged carbon through their bodies and release methane, where it reacts in the atmosphere to form CO_2.) In a natural scenario, without microbes, there wouldn't be enough CO_2 in the atmosphere to balance the oxygen produced by photosynthesizers, and if enough oxygen accumulated in the atmosphere, the sky would catch fire. Some scientists have even proposed the presence of methane on Mars may be the result of native Martian methanogens.

Biology, wrote the microbiologist John Ingraham in *March of the Microbes*, creates Earth's chemistry. All the major chemical cycles on our planet have a microbial component: carbon, nitrogen, oxygen, phosphorus, sulfur, water, even metals. Bacteria living deep in Earth's crust secrete uranium. It's produced biologically over millions of years. This has led to a "paradigm shift in ore genesis," said the coauthor of the paper on the discovery. Indeed, the more we learn about microbes, the greater the paradigm shift in our understanding of nature.

But have the algae evolved to help themselves by seeding clouds that cool the surface layers of the ocean, since temperatures above 53°F kill off surface algae? Do the fungi in the Pacific Northwest rainforests produce lighter-than-air spores to seed clouds that drop the rain that supports their growth? "This makes no more sense than saying animals exhale CO_2 in order to support plant productivity," wrote the mycologist Nicholas Money in *The Amoeba in the Room*. It is unlikely there is deliberate manipulation of the atmosphere by microbes. There's no evolutionary goal on their part. Nature isn't teleological. It's adaptive. If seeding a cloud caused bacteria on the ground to die, then eventually only those bacteria that didn't seed clouds would be the ones to go on to survive.

Conditions elsewhere in the system matter, however. Fungal spores may seed clouds, but not if there is no cow manure for the fungi to grow in, so the well-being of the cow would be important to the seeding of the cloud and the fulfillment of the cycle. And, of course, there's no stopping there. The cow needs grass, the grass needs nitrogen, and on and on. "It's all interconnected," said Dr. Sands.

The pulse of these systems, where biota (life) influence abiota (nonlife) and back again, is a negative feedback loop. A negative feedback loop tends to reduce fluctuations. (This drove me crazy in biology class because a negative

feedback loop leads to stability whereas a positive feedback loop leads to insta-
bility.) The sun heats the ocean, the algae make DMSP, the DMSP converts into
DMS, the DMS seeds the clouds, the clouds cool the ocean. Negative feedback
loops regulate environmental stability; they maintain a status quo. The same is
true inside our bodies. The feedback loops between gut microbes and our
immune system, for example, regulate inflammation. Environmental stability
is an emergent property of all the interactions between life and life, and life and
nonlife. It's not a miracle, but it feels like one.

The Englishman Dr. James Lovelock (he was 97 at the time of this writing),
a chemist and inventor, imagined the entire plant was autopoietic (from the
Greek for "self" and "creation"). He crafted the Gaia hypothesis, which
the biologist Lynn Margulis defined in her book *Symbiotic Planet* as "the series
of interacting ecosystems that compose a single huge ecosystem at
the Earth's surface."* In the 1960s, Dr. Lovelock was working for NASA and got
the idea looking at the atmospheres of Mars and Venus. They have stable chemical equilibriums. There is no sign of change over time. They are still. The
Earth, however, is in a state of extreme disequilibrium. Things are changing all
the time. Dr. Lovelock and his colleagues determined that Earth might be a sort
of superorganism, where the atmosphere, oceans, climate, and crust are constantly being regulated in a way that is comfortable for life by the behavior of
living organisms. "I see the Earth as more than just a mixture of living things
and inanimate matter," said Dr. Lovelock in an interview. "I see it as a tightly
coupled entity, where the evolution of the living things and the evolution of the
inorganic matter constitute a single and totally inseparable process. It's a whole
system."

This blew the minds of many people. The New Age community in the
1970s tapped not into the Gaia hypothesis, which describes an unconscious
system (as Lynn Margulis put it, the Earth, if it was sentient, would be about as
sentient as an amoeba), but rather into the "Spirit of Gaia," a kind of world soul,
according to David Spangler, one of the New Age movement's most well-known
spokespersons. And certainly, the New Agers' interest didn't soften the stance

---

*The novelist William Golding proposed the name Gaia for the hypothesis. Gaia (pronounced
guy-a) is the name of the Greek goddess of the Earth. She's a member of the Protogenoi, primeval
gods that also include Chaos, Nature, Day, Night, and so on.

of the scientists who were skeptical that natural selection operating on individual organisms could lead to planet-wide stability. In response, Dr. Lovelock developed a mathematical model called Daisyworld. It illustrates how feedback mechanisms can evolve from the actions of self-interested organisms, rather than natural selection.

Much of the scientific community thinks Gaia is a farfetched idea, though in 2001, scientists from a collection of international global change research programs produced the "Amsterdam Declaration on Earth System Science," which stated, "The Earth System behaves as a single, self-regulating system comprised of physical, chemical, biological and human components." It's not a full-throated endorsement of Gaia, but it seems to be getting to the same notion—that all parts of the planet, living and not living, are connected.

At the end of the exam for his graduate course, Integrative Microbiology, Moselio Schaechter offered his students an extra-credit question. How would you go about defending the statement that "all living things are connected to other living things" to an educated lay audience? The student answers ran the gamut, but I liked this one best.

"There may be just one living thing, life, with many working components."

After graduation, though it wasn't officially my graduation and so barely qualified as a congratulatory present to myself, I bought a microscope. It's a beauty. A 40X to 1000X Research Compound Binocular Microscope made of black and enameled metal, with focusing knobs that move smoothly and slowly and silently, as if through syrup. But it sat on my desk for months. I never took off the plastic hood because I was afraid of it. I'm not a scientist.

And then, about a year after I returned from Boston, I went to a microscopy class that my mushroom club sponsored. The class was focused on seeing a fungal spore. I could hear crows of delight as my fellow students pulled focus on their microscopes and the invisible world became visible. "Shall I look with glasses or without glasses?" asked the woman sitting next to me. We fussed with the cover slips over slides, we dropped overly large drops of ammonia hydroxide onto our samples, we chased the floaters on our eyeballs. We were looking at 400X magnification. I'd need 1000X to see bacteria. And I wanted to see mine.

I contacted my former TA Kyle and asked for his help. Kyle brought by some methylene blue stain, which we would need to see my cells since they are naturally transparent. He warned me, "Everything this stain touches will turn blue," and so I geared up in rubber dishwashing gloves. He talked me through it, admiring my new microscope, a martini at his elbow (he asked if I minded cocktails on my bench, a bit of lab jargon that gave me a science-groupie thrill). I added a drop of water to the slide and then used a dry toothbrush to rub some cells from the inside of my cheek that I tapped onto the glass. I added a drop of water, the tiniest bit of methylene blue, the cover slide, and a smidge of mineral oil, and looked.

I pulled focus in and out, moved the slide a bit to the left, to the right, tried to see with my glasses, tried to see without. Sat back, took a breather, leaned in again, pulled focus, first on some mucky stuff that I couldn't identify. Then I spotted a couple of my cells, misshaped goopy-looking things with the nucleus like a dark, cloudy eye. I adjusted the focus with the lightest touch I could.

And then I saw them. Dozens of black balls tumbling through space like astronauts whose lifelines were cut. When I focused on one bacterium, I could see it tumble and then travel straight in one direction for a distance maybe 100 times the length of its body, then stop and sense. And then the tumbling would begin again, and the bacterium headed in another direction. They were hunting, not with senses like ours, not with eyes or ears, but with chemical receptors, sampling their miniscule environment for signs of food, a fat molecule here, a protein molecule there. They were living their lives in their real time, parallel to me as I lived mine, both of us components of one living spectrum. "Hello," I kept saying, "hello." Kyle smiled and took a sip of his drink. "If you like that, you should try a drop of water from one of the puddles in the street sometime," he said. "After it rains, a whole ecology blossoms."

I was awed. It was a feeling that fell at the interface of the amazing truths science reveals and a deep intuition that those truths are transcendent. When I first started to study biology, I explained to my daughter, Carson, that I wanted to understand the deep feeling of connectedness with nature and other people that I so frequently had, what the ecologist Stan Rowe described as "the sense of wonder and affection felt for the splendor and bounty of the Earth."

"It sounds like you are just trying to find a scientific answer for your spiritual feelings," she said.

I had to laugh because she was right. I wasn't looking to find out microbes are God (though I am tempted; it kind of makes sense that God wouldn't be one great huge entity, but actually billions and billions of tiny ones), but I did learn that microbes are the basis of all things biological, that they created living, and to a great degree, do our living on our behalf. They are our ancestors, and our most intimate companions, and they permeate and link all life on Earth. It took a glimpse into the unseen world for me to see what I already knew intuitively. We have coevolved with all that lives on the planet, and as a result, we share *everything*.

I shut off the light and cleaned my slides, and I pushed my microscope to the end of the desk. But I didn't cover it with its plastic hood.

I had a whole world to look at tomorrow.

Acknowledgments

I took on this book because I didn't know any better. When I explained what I was doing quite a few people wished me luck, kind of the way you might wish someone luck who is planning to represent themselves in court. Nonetheless, many very smart, generous folks were willing to help me wade through this daunting material.

I am very grateful to my professors at Columbia: Shahid Naeem, Dustin Rubenstein, Paul Olsen, Natalie Boelman, and Matthew Palmer. Early drafts of the book were read by Kyle Frischkorn, my lab teacher and a PhD candidate in microbiology at the Lamont-Doherty Earth Observatory, and Jonathan Rosenthal, who is in premed at NYU. They helped me get the basic science right, and I gratefully incorporated their corrections. Once the manuscript was done, the microbiologists Moselio Schaechter and Phil Strandwitz read it through, and their observations helped refine the text. Many thanks to Justin Kavanagh, whose insightful edits improved the book in countless ways. I can't thank my lay readers Ezra Herman and Paul Sadowski enough. They spent an inordinate amount of time and energy, considering every statement carefully, a task which I can only chalk up to love. Thanks to Pam Krauss for her insights. She's tough, which is only one of the reasons why I admire her.

Three scientists were especially gracious about reading parts of the manuscript that pertain to their area of expertise. Timothy Crews, director of research and lead scientist at the Land Institute, read my chapters on soil and plants for accuracy. James White, mycologist and plant pathologist at Rutgers University, also read the chapters about soil and plants, and more. He was the godfather of this book, spending hours talking to me about microbes over beer and fish and chips. Congratulations on having *Jimwhitea*, a genus of Triassic fungi named after you, Jim! Kael Fischer, CEO at uBiota, read the chapters on the human microbiome. My first interview with Kael was so over my head that I had to take a nap afterward to recover. But as time went on Kael found my level and was consistently merciful and available and supportive. Thanks to astronomer Deno Stelter for reviewing my chapter on the origins of the universe.

I interviewed many scientists in the process of writing this book. I'm thankful for the hours I was fortunate enough to spend in conversation with the brilliant Nicholas Money, mycologist at the University of Miami, Ohio. Many thanks to Kristine Nichols, chief scientist at the Rodale Institute; Russell Rodriguez and Regina Redman of Adaptive Symbiotic Technologies; Tom Volk, mycologist at the University of Wisconsin-La Crosse; and Mary Evelyn Tucker, of Yale University. A conversation with Mary Evelyn is like a flying dream: it soars. Thank you, Krista McGuire, biologist at Barnard College; microbiologist Scott Clark; Susan Holmes of Lane Community College in Eugene, Oregon; and Brendan Mormile, a PhD candidate at Texas A&M whose work on fungal endophyte/bacterial endophyte interactions in fescue grass is so elegant. Thanks to Britt Bunyard of *Fungi Magazine*, and Camille English, PhD candidate at Rutgers University, for letting me email you about a million times; and Douglas Eveleigh of Rutgers University, for the speed-date history lesson in microbiology.

Most of the scientists and academics I spoke to have or will have doctorates by the time you read this. I chose not to add their honorifics in the text but it should be noted they have devoted many years of study to their respective fields and through their ongoing research, publications, and lectures, they continue to add to our understanding of the unseen world.

I didn't talk exclusively with scientists but also farmers and agriculture professionals. I am especially grateful for the time and hospitality of Ron Heck; the Sweeney family; James Gillespie, Division of Soil Conservation & Water Quality, Iowa Department of Agriculture; David Ertl, Iowa Corn Promotion Board; Don Smith from Kiss the Ground; Lance Hansen of Jack Rabbit Hill Winery in Hotchkiss, Colorado; and the artist Andrew Cziraki.

I hope I have done your efforts justice. Forgive me for any errors. They are indisputably mine.

Many thanks to family and friends who shared quotes: Jasper Hartnett; Neni Panourgia; Tim Grabham; Paul Guilfoyle, former rust-eater; Sean Sullivan; Gary Lincoff, whose generosity of spirit and mind continues to inspire me; Carson Bone, ever insightful; Mo Bone, always hilarious; and Edward Giobbi, for his poops. Thank you to my many friends and colleagues who sent me articles, scientific papers, video clips, and images of microbes. Yes, Al Appleton, I read them all.

I have the annoying habit of working through ideas in dinner table conversation. You know who your friends are when they will sit through an hour of someone struggling to describe what a bacteriophage is and they still agree to hang out with you the following week. Among the most stalwart are Julie Rigby, Jody Guralnick, Michael Lipkin, Lisa Giobbi, Alex Neil, John Vennema, Diane and Gerard Koeppel, Cham Giobbi, Laine Valentino, Rachel Stroer from the Land Institute, James Oseland, Susan Murrmann, Nathalie Smith, John Zito, Jayne Morrell, Richard Zacks, Kristine Dahl, Louis Schwartzberg, and Stathis Gourgouris. Thanks for putting up with me Cornelia Cho, Sam Landes, Steve and Linda Rubick, Katherine McCarthy, Liz Lilien, Candy and Bob Penetta, Marley and Linda Hodgson, and Barb and Mike Heck. I am grateful for my sisters at Downtown Women for Change and my friends in the New York Mycological Society, and mushroom clubs and enthusiasts further afield, but especially David Campbell. Thank you Susan Goldhor and Aron Bernstein for your generosity and friendship, and special thanks to Susan, who read my manuscript and warned me about sounding whiney. Oh, and Michael Simmons. Sorry about the Starbucks thing.

On the bookmaking side, thanks to my agent, Angela Miller. I am usually low maintenance, but this time around I think I was a pretty exhausting client. Thank you, Angela, for being the ally I needed. Thank you to the Rodale team, Leah Miller, who took the leap of faith necessary to get this book started, Anna Cooperberg, Dervla Kelly, and Jennifer Levesque, the excellent Michelle Janowitz, Kathy Dvorsky, art director Amy King, Brianne Sperber, Marilyn Hauptly, Gail Gonzales, and the perfect, poised, and persistent Aly Mostel.

I am sure I have forgotten people who helped me over the years. For that oversight, please forgive me.

Most especially, I want to thank Kevin Bone. During the years it took to write this book, all the things I am in the habit of doing to help keep our lives running smoothly were chucked out the window. Kevin, I'm sorry I bought those cookbooks and tried to make it look like I was supporting your suppressed passion for cooking when actually I had abandoned my part of our long-held division of labor. For all that, and more, including your excellent read of the manuscript, thank you. Every day I am grateful that my microbes are homogenized with yours.

Notes

Introduction

Much of the Introduction is first-person narrative combined from material found elsewhere in the book. Attributions for that material can be found in the Notes where they appear in the chapters. But for general biology, most of what I report I learned in lectures by Shahid Naeem, Dustin Rubenstein, Natalie Boelman, Paul Olsen (whose section I do not reference in the book), and Matthew Palmer during the two-semester course Environmental Biology I and II and in our textbook, *Biology* by Kenneth Mason et al. (New York: McGraw-Hill, 2014). Because the class didn't dwell on microbiology, I often returned to these two books by Lynn Margulis and Dorion Sagan: *Microcosmos: Four Billion Years of Microbial Evolution* (Berkeley: University of California Press, 1997) and *Garden of Microbial Delights: A Practical Guide to the Subvisible World* (Dubuque, IA: Kendall Hunt, 1993). Also very helpful was Moselio Schaechter, John L. Ingraham, and Frederick C. Neidhardt, *Microbe* (Washington, DC: ASM Press, 2006); Carl Zimmer, *Microcosm: E. coli and the New Science of Life* (New York: Vintage, 2009); Elio Schaechter (yes, same guy as Moselio), *In the Company of Microbes: Ten Years of Small Things Considered* (Washington, DC: ASM Press, 2016); Roberto Kolter and Stanley Maloy, eds., *Microbes and Evolution: The World That Darwin Never Saw* (Washington, DC: ASM Press, 2012); Nicholas P. Money, *The Amoeba in the Room: Lives of the Microbes* (Oxford, England: Oxford University Press, 2014); and Lewis Thomas, *The Lives of a Cell: Notes of a Biology Watcher* (New York: Viking, 1974), who proves that, yes, microbiology can be lyrical.

In the Introduction, I quoted Tom Volk from my book *Mycophilia: Revelations from the Weird World of Mushrooms* (Emmaus, PA: Rodale, 2011) and Stephen Jay Gould, "The Evolution of Life on the Earth," *Scientific American*, October 1994. Tom Curtis's quote comes from *Nature Reviews Microbiology* 4 (2006): 488, nature.com /articles/nrmicro1455. Andrew Cziraki's The Holy Water Project was supported by the Bio Art Lab at the School of Visual Arts in New York City.

Chapter 1

The microbiology blog *Small Things Considered* can be found at schaechter .asmblog.org. It can get pretty technical, which is why I signed up for Biology 101 at Universal Class: universalclass.com/i/course/biology-101.htm.

Barnard College announced its transgender admissions policy in 2015: barnard .edu/news/barnard-announces-transgender-admissions-policy.

There are numerous websites about going back to college. A few I read are back2college.com, nces.ed.gov, and forums.welltrainedmind.com.

The textbook I tried to read over the summer was *Campbell Biology* by Jane B. Reece et al. (Boston: Benjamin Cummings, 2011). Unfortunately, not the same book I'd be obliged to purchase once school started.

The research on adolescent digital multitasking is scant: Sarayu Caulfield and Alexandra Ulmer, "Capacity Limits of Working Memory: The Impact of Media Multitasking on Cognitive Control and Emotion Recognition in the Adolescent Mind," aap.confex.com/aap/2014/webprogram/Paper27323.html, but anecdotally, if your kids are digital natives, you've seen them in action.

Additional random information about the origins of the universe came from a variety of online sources, including lecture notes on the "Evolution of the Atmosphere" by Perry Samson from the University of Michigan found at globalchange.umich.edu /globalchange1/current/lectures/Perry_Samson_lectures/evolution_atm. I regularly visited scientificamerican.com, nationalgeographic.com, phys.org, rosetta.jpl.nasa.gov /rosetta-science-blog, and askamathematician.com. I also learned a lot from enlightening email exchanges with the astronomer Deno Stelter of the University of Florida.

There are different hypotheses for the origin of life. To report on some of these, I referred to Nick Lane, *The Vital Question: Energy, Evolution, and the Origins of Complex Life* (New York: W.W. Norton & Company, 2015); Clair Edwin Folsome, *The Origin of Life: A Warm Little Pond* (San Francisco: W. H. Freeman and Company, 1979); Richard Fortey, *Life: A Natural History of the First Four Billion Years of Life on Earth* (New York: Vintage, 1999); and Geoffrey M. Cooper, *The Cell: A Molecular Approach*, 2nd edition (Sunderland, MA: Sinauer Associates, 2000). I also used information from the following sources: Ashutosh Jogalekar, "The Beginnings of Life: Chemistry's Grand Question," *The Curious Wavefunction* (blog), *Scientific American*, 2012, blogs .scientificamerican.com/the-curious-wavefunction/the-beginnings-of-life-chemistrys -grand-question; Hideo Hashizume, "Role of Clay Minerals in Chemical Evolution and the Origins of Life," in *Clay Minerals in Nature—Their Characterization, Modification and Application*, eds. Marta Valaškova and Grażyna Simha Martynkova (InTech, 2012), doi.org/10.5772/2708; Gemma Reguera, "Microbial 'Starstuff,'" *Small Things Considered* (blog), 2013, schaechter.asmblog.org/schaechter/2013/07/microbial -starstuff.html; Robert F. Service, "Researchers May Have Solved Origin-of-Life Conundrum," *Science*, March 16, 2015, sciencemag.org/news/2015/03/researchers -may-have-solved-origin-life-conundrum; and the video "Revealing the Origins of Life," *NOVA scienceNOW*, aired 2011, on PBS, naturedocumentaries.org/3023/origin -life-pbs-2011.

I quoted Stanley L. Miller in "From Primordial Soup to the Prebiotic Beach," an interview by Sean Henahan, Contra Costa College, October 1996, cccbiotechnology .com/WN/NM/miller.php. Carl Sagan's quotes are jewels. I picked the one I used on goodreads.com/work/quotes/3237312-cosmos. It is from his book *Cosmos,* which sold almost as many copies as stars in the sky (New York: Ballantine, 1980).

I pulled information on the Tagish Lake meteorite from "The Tagish Lake Meteorite," *Astrobiology Magazine*, June 2, 2002, astrobio.net/news-exclusive/the-tagish-lake-meteorite and on the Allan Hills meteorite from "Allan Hills Meteorite Abiogenic?," *Astrobiology Magazine*, July 21, 2004, astrobio.net/mars/allan-hills -meteorite-abiogenic and from Bill Steigerwald, "Asteroid Served Up 'Custom Orders' of Life's Ingredients," Goddard Space Flight Center, June 9, 2011, nasa.gov/centers /goddard/news/features/2011/tagish-lake.html.

For information about ancient ribocytes, I referred to Marcia Stone, "Oddly Microbial: Ribocytes," *Small Thing Considered* (blog), 2012, schaechter.asmblog.org /schaechter/2012/04/oddly-microbial.html and Michael Yarus, "Primordial Genetics: Phenotype of the Ribocyte," *Annual Review of Genetics* 36 (2002): 125–51. My tutor Jonathan Rosenthal gave me the analogy that DNA is like an encyclopedia and RNA is like the "how to build a bicycle" entry.

Once macromolecules became animated, the science gets a lot less murky. I learned about membranes in class and from my biology book.

Chapter 2

To report on the book bag weight problem, I read numerous articles by S. Dockrell, C. Simms, and C. Blake, including "Guidelines for Schoolbag Carriage: An Appraisal of Safe Load Limits for Schoolbag Weight and Duration of Carriage," *Work* 53, no. 3 (2015): 679–88, and reporting by WALB News, accessed August 13, 2017, walb.com /story/12934127/study-heavy-backpacks-related-to-spine-problems.

Additional information on what early Earth was like was gleaned from *Life* by Richard Fortey; "Earth's Early Atmosphere," *Astrobiology Magazine*, December 2, 2011, astrobio.net/geology/earths-early-atmosphere; and "The Earliest Atmosphere," Smithsonian Environmental Research Center, forces.si.edu/atmosphere/02_02_01.html.

I learned about what swimming is like for a bacterium from Madison Krieger, "Bacteria Are Masters of Tai Chi: The Remarkable Science That Helped Me Understand What It Means to Be a Physicist," *Nautilus*, June 2, 2016, nautil.us/issue/37/currents /bacteria-are-masters-of-tai-chi.

To build this chapter, I referred to many sources describing microbes, but I benefitted most from descriptions in *Garden of Microbial Delights* by Lynn Margulis and Dorion Sagan, *Deadly Companions: How Microbes Shaped Our History* by Dorothy H. Crawford (New York: Oxford University Press, 2007), and *Microbe* by Moselio Schaechter et al. You don't have to look far to find headlines about how dirty (as in bacteria-infected) any household item is, but I learned about the state of my keyboard on abcnews.com, accessed August 13, 2017, abcnews.go.com/Health/Germs /story?id=4774746. I read the analogy regarding the size of a bacterium relative to us in the terrific book by David R. Montgomery and Anne Biklé, *The Hidden Half of Nature: The Microbial Roots of Life and Health* (New York: W.W. Norton & Company, 2016). My lab TA at Columbia, Kyle Frischkorn, described the prokaryote's pilus as a penis. Marisa Pedulla's quote about bacteriophages, the viruses that predate bacteria, was in the article by John Travis, "All the World's a Phage: Viruses That Eat Bacteria Abound—and

Surprise," *Science News,* July 12, 2003, phschool.com/science/science_news/articles /all_worlds_phage.html. For a deeper understanding of CRISPR, I read Michael Specter, "Rewriting the Code of Life," *New Yorker,* January 2, 2017, newyorker.com /magazine/2017/01/02/rewriting-the-code-of-life.

Prokaryotes grow exponentially. To describe this growth, I referred to examples in *Deadly Companions* by Dorothy H. Crawford; *Microcosm* by Carl Zimmer; Jeff Lowenfels and Wayne Lewis, *Teaming with Microbes: The Organic Gardener's Guide to the Soil Food Web* (Portland, OR: Timber Press, 2010); and Roberto Kolter, "Fine Reading: No Growth!," *Small Things Considered* (blog), 2017, schaechter.asmblog.org/schaechter /2017/07/fine-reading-no-growth.html?utm_source=feedburner&utm_medium=feed &utm_campaign=Feed%3A+schaechter+%28Small+Things+Considered%29. Prokaryotes can share beneficial genes at an incredible rate, as per the following articles: Oliver Tenaillon et al., "Tempo and Mode of Genome Evolution in a 50,000-Generation Experiment," *Nature* 536 (2016): 165–70 and Rene Niehus et al., "Migration and Horizontal Gene Transfer Divide Microbial Genomes into Multiple Niches," *Nature Communications* 6 (2015): doi.org/10.1038/ncomms9924.

One bacterium can't do very much, but a biofilm can do a lot. I built this section from many sources, but I leaned most heavily on: Rodney M. Donlan, "Biofilms: Microbial Life on Surfaces," *Emerging Infectious Diseases* 8, no. 9 (2002): 881–90; J. W. Costerton, L. Montanaro, and C. R. Arciola, "Biofilm in Implant Infections: Its Production and Regulation," *International Journal of Artificial Organs* 28, no. 11 (2005): 1062–68; Marianne Spoon, "How Biofilms Work," HowStuffWorks, accessed August 13, 2017, science.howstuffworks.com/life/cellular-microscopic/biofilm.htm; and Alfred B. Cunningham, John E. Lennox, and Rockford J. Ross, eds., *Biofilms: The Hypertextbook* (2001–2008), cs.montana.edu/webworks/projects/stevesbook/contents/appendices /appendix001/pages/page002.html. The article "Understanding Biofilms," Bacteriality.com, May 26, 2008, accessed August 13, 2017, bacteriality.com/2008/05/biofilm/ was especially awesome.

Quorum sensing is the ability of bacteria to sense their numbers. Once they reach a certain number, bacteria can form a biofilm and do all kinds of things they weren't able to do as a single bacterium. I found the following articles very helpful in reporting on this new and exciting area of research: R. Fredrik Inglis et al., "Spite and Virulence in the Bacterium *Pseudomonas aeruginosa,*" *PNAS* 106, no. 14 (2009): 5703–707; "What Is Quorum Sensing?," The Quorum Sensing Site, accessed August 13, 2017, nottingham.ac.uk /quorum/what.htm; Ada Hagan, "Kiss and Make Up: *Myxococcus xanthus* Demonstrates Bacterial Cooperation," *Small Things Considered* (blog), 2015, schaechter.asmblog .org/schaechter/2015/10/kiss-and-make-up-myxococcus-xanthus-demonstrates-bacterial -cooperation.html; Pamela Lyon, "From Quorum to Cooperation: Lessons from Bacterial Sociality for Evolutionary Theory," *Studies in History and Philosophy of Science Part C: Studies in History and Philosophy of Biological and Biomedical Sciences* 38, no. 4 (2007): 820–33; Samay Pande et al., "Privatization of Cooperative Benefits Stabilizes Mutualistic Cross-Feeding Interactions in Spatially Structured Environments," *ISME*

Journal 10, no. 6 (2016): 1413–23; "Biologists Discover Timesharing Strategy in Bacteria: Communities Found to Coordinate Feeding to Streamline Efficiency," ScienceDaily, April 6, 2017, accessed August 13, 2017, sciencedaily.com/releases/2017/04/170406143912 .htm; Elio Schaechter, "Coping with Hard Times: Death as an Option," *Small Things Considered* (blog), 2011, schaechter.asmblog.org/schaechter/2011/04/coping-with-hard -times-death-as-an-option.html?utm_source=feedburner&utm_medium=email&utm _campaign=Feed%3A+schaechter+%28Small+Things+Considered%29; and Michael E. Hibbing et al., "Bacterial Competition: Surviving and Thriving in the Microbial Jungle," *Nature Reviews Microbiology* 8 (2010): 15–25 (a fantastic article that equates bacteria communities with human ones). But it all came together when I heard a talk on quorum sensing by Bonnie L. Bassler from Princeton University, who spoke at the 2016 Lynford Lecture at NYU's Tandon School of Engineering on October 12, 2016. Wow.

To describe endospores, I read "Bacterial Endospores," Department of Microbiology, Cornell College of Agriculture and Life Sciences, accessed August 17, 2017, micro .cornell.edu/research/epulopiscium/bacterial-endospores and Wayne L. Nicholson et al., "Resistance of Bacillus Endospores to Extreme Terrestrial and Extraterrestrial Environments," *Microbiology and Molecular Biology Reviews* 64, no. 3 (2000): 548–72, which was out of this world.

I reported the section on the origin of the nucleus from "The Eukaryotic Nucleus," accessed August 17, 2017, bio.sunyorange.edu/updated2/GENETICS/10%20nucleus .htm. The kooky human-to-bacteria weight analogy comes from "What Weighs More— Bacteria or Every Single Person on Earth?," uBiome blog, 2016, ubiomeblog.com /weighs-bacteria-every-single-person-earth. The reference to microbial dark matter is drawn from E.O. Wilson, "Exploring the Complexity of Life" Convocation, Life Science Institute, University of Michigan, May 14, 2004, lsi.umich.edu/eo-wilson-exploring- the-complexity-of-life.

I used the following sources to describe extremophiles: John L. Ingraham, *March of the Microbes: Sighting the Unseen* (Cambridge, MA: Harvard University Press, 2012); *Microbe* by Moselio Schaechter et al.; Thomas D. Brock, *Life at High Temperatures* (Yellowstone Association, 1994); and *Garden of Microbial Delights* by Lynn Margulis and Dorion Sagan. Beyond these sources, I also used Nicholas Wade, "Meet Luca, the Ancestor of All Living Things," *New York Times,* July 25, 2016; R. Cavicchioli, "Extremophiles and the Search for Extraterrestrial Life," *Astrobiology* 2, no. 3 (2002): 281–92; Nola Taylor Redd, "Hardy Bacteria Thrive Under Hot Desert Rocks," *Astrobiology Magazine,* June 1, 2015, astrobio.net/mars/hardy-bacteria-thrive-under-hot-desert -rocks; Ed Yong, "Life Found Deep under the Sea," *Nature,* March 14, 2013, nature.com /news/life-found-deep-under-the-sea-1.12610; Lynn J. Rothschild and Rocco L. Mancinelli, "Life in Extreme Environments," *Nature* 409 (2001): 1092–101; Peter Tyson, "The Lives of Extremophiles," NOVA, October 1, 2002, accessed August 13, 2017, pbs.org /wgbh/nova/nature/lives-of-extremophiles.html; Brett J. Baker et al., "Enigmatic, Ultrasmall, Uncultivated Archaea," *PNAS* 107, no. 19 (2010): 8806–11; Kristian Sjøgren, "Live Bacteria Found Deep Below the Seabed," ScienceNordic, accessed August 17, 2017, sciencenordic.com/live-bacteria-found-deep-below-seabed; Carrie Arnold, "Pushing the Limits of Life," NOVA Next, November 4, 2015, accessed August 17, 2017, pbs.org /wgbh/nova/next/earth/deep-life; "Space Station Research Shows That Hardy Little Space Travelers Could Colonize Mars," nasa.gov, May 2, 2014, accessed January 10, 2018, nasa.gov/mission_pages/station/research/news/en_tef/; and Daniel Cressey, "Genetic Popsicle," *Nature*, August 7, 2007, nature.com/news/2007/070806/full /news070806-4.html.

I learned about photosynthesis from endless descriptions in class and mind-boggling diagrams in my biology book, but I ended up using descriptions from John Archibald, *One Plus One Equals One: Symbiosis and the Evolution of Complex Life*, probably the best title of a biology book *ever* (Oxford, England: Oxford University Press, 2014). Paul G. Falkowski, *Life's Engine: How Microbes Made Earth Habitable* (Princeton, NJ: Princeton University Press, 2015) was also a great resource.

The story of iron is a fascinating one, and most of what I learned, I learned from interviews with Moselio Schaechter and a variety of papers, including "The Mechanisms of Iron Absorption," Information Center for Sickle Cell and Thalassemic Disorders, accessed August 17, 2017, sickle.bwh.harvard.edu/iron_absorption.html; Stu Borman, "Novel Vaccines Block Iron Scavenging by Bacteria," *Chemical & Engineering News* 94, no. 45 (2016): 7; and P. T. Lieu et al., "The Roles of Iron in Health and Disease," *Molecular Aspects of Medicine* 22, no. 1–2 (2001): 1–87.

Regarding the oxygenation of the atmosphere, I used many sources, including "The Rise of Oxygen," American Museum of Natural History, accessed August 17, 2017, amnh.org/explore/science-bulletins/(watch)/earth/documentaries/the-rise-of-oxygen/ and John M. Kennard and Noel P. James, "Thrombolites and Stromatolites: Two Distinct Types of Microbial Structures," *PALAIOS* 1 (1986): 492–503.

Chapter 3

There are many studies on the subject of GPA and classroom seating preferences. The most useful papers I read were Katherine K. Perkins and Carl E. Wieman, "The Surprising Impact of Seat Location on Student Performance," *Physics Teacher* 43 (2005): 30–33 and Michael D. Meeks et al., "The Impact of Seating Location and Seating Type on Student Performance," *Education Sciences* 3 (2013): 375–86. In my search for guidance on note-taking, I read Françoise Boch and Annie Piolat, "Note Taking and Learning: A Summary of Research," *WAC Journal* 16 (2005): 101–12, but I got the most out of the various note-taking tip pages on the internet, like Thorin Klosowski, "Back to Basics: Perfect Your Note-Taking Techniques," Lifehacker, accessed August 17, 2017, lifehacker.com/back-to-basics-perfect-your-note-taking-techniques-484879924 and Laura Pappano, "How to Take Better Lecture Notes," *New York Times*, October 31, 2014.

I read numerous books and articles about classification of organisms, including James Lennox, "Aristotle's Biology," in *Stanford Encyclopedia of Philosophy*, ed. Edward N. Zalta (Metaphysics Research Lab, Stanford University, 2017), plato.stanford.edu /archives/spr2017/entries/aristotle-biology; Lois Tilton, "From Aristotle to Linnaeus:

The History of Taxonomy," *Dave's Garden* (blog), 2009, davesgarden.com/guides /articles/view/2051/#b; Harriet Ritvo, *The Platypus and the Mermaid and Other Figments of the Classifying Imagination* (Cambridge, MA: Harvard University Press, 1997); T. H. White, ed., *The Book of Beasts* (New York: Putnam, 1960); Ruth A. Johnston, *All Things Medieval: An Encyclopedia of the Medieval World, Volume 1* (Santa Barbara, CA: Greenwood, 2011); and Michel Foucault, *The Order of Things: An Archaeology of the Human Sciences* (New York: Vintage, 1973). I fell down numerous time-consuming wormholes about curious creatures in zoology when I discovered the biodiversitylibrary.org.

To understand microbial taxonomy, I used the following: John G. Simmons, *Doctors and Discoveries: Lives That Created Today's Medicine* (Boston: Houghton Mifflin, 2002); Marc J. Ratcliff, *The Quest for the Invisible: Microscopy in the Enlightenment* (London: Routledge, 2009), Kindle edition; Lois N. Magner, *A History of Infectious Diseases and the Microbial World* (Westport, CT: Praeger, 2009); Lynn Margulis, *Symbiotic Planet: A New Look at Evolution* (New York: Basic Books, 1998); W. Ford Doolittle and R. Thane Papke, "Genomics and the Bacterial Species Problem," *Genome Biology* 7, no. 9 (2006): 116; Rachel J. Whitaker, "A New Age of Naturalists," *Microbe* 6, no. 11 (2011): 491–94; the website historyofvaccines.org; Darren R. Flower, *Bioinformatics for Vaccinology* (Hoboken, NJ: John Wiley & Sons, 2009); O. Assadian and G. Stanek, "Theobald Smith—The Discoverer of Ticks as Vectors of Disease," *Wiener klinische Wochenschrift* 114, no. 13–14 (2002): 479–81; R. G. E. Murray and John G. Holt, "The History of Bergey's Manual," in *Bergey's Manual of Systematic Bacteriology*, eds. D. J. Brenner et al. (Boston: Springer, 2005); P. G. Heineman, "Orla-Jensen's Classification of Lactic Acid Bacteria," *Journal of Dairy Science* 3, no. 2 (1920): 143–55; Edward Shorter, "Ignaz Semmelweis: The Etiology, Concept, and Prophylaxis of Childbed Fever," *Medical History* 28, no. 3 (1984): 334; Noha H. Youssef et al., "Assessing the Global Phylum Level Diversity within the Bacterial Domain: A Review," *Journal of Advanced Research* 6, no. 3 (2015): 269–82; the Tree of Life web project, tolweb.org, accessed August 17, 2017; a feisty 10-minute lecture from Douglas Eveleigh at the beery Bactoberfest at Rutgers University in October 2016; and the exquisite *Microbe Hunters* by Paul de Kruif (New York: Harcourt Brace, 1926).

Newfound groups of bacteria are not making classification any easier. I used the following to try and sort it out: Laura A. Hug et al., "A New View of the Tree of Life," *Nature Microbiology* 16048 (2016): doi.org/10.1038/nmicrobiol.2016.48 and Kevin Hartnett, "At Tiny Scales, a Giant Burst on Tree of Life," *Quanta Magazine*, July 28, 2015, quantamagazine.org/newfound-bacteria-expand-tree-of-life-20150728. Additionally, John D. Sutherland, "Opinion: Studies on the Origin of Life—The End of the Beginning," *Nature Reviews Chemistry* (2017): doi.org/10.1038/s41570-016-0012; William A. Rosche and Patricia L. Foster, "Determining Mutation Rates in Bacterial Populations," *Methods* 20, no. 1 (2000): 4–17; J. L. Martinez and F. Baquero, "Mutation Frequencies and Antibiotic Resistance," *Antimicrobial Agents and Chemotherapy* 44, no. 7 (2000): 1771–77; Michael S. Rappé and Stephen J. Giovannoni, "The Uncultured Microbial Majority," *Annual Review of Microbiology* 57 (2003): 369–94; Stephen J. Giovannoni, J. Cameron Thrash, and Ben Temperton, "Implications of Streamlining Theory for

Microbial Ecology," *ISME Journal* 8 (2014): 1553–65, doi.org/10.1038/ismej.2014.60; and K. J. Locey and Jay T. Lennon, "Scaling Laws Predict Global Microbial Diversity," *PNAS* 113, no. 21 (2016): 5970–75. A video from Harvard Medical School—"The Evolution of Bacteria on a 'Mega-Plate' Petri Dish," YouTube video, 1:54, posted by Kishony Lab, 2016—illustrates the mutational speed of bacteria like nothing else. You've got to see it to believe it: youtube.com/watch?v=plVk4NVIUh8.

About particular archaeal and bacterial taxa, I beefed up my reporting with information gleaned from *Microbe* by Moselio Schaechter et al.; Jacquelyn G. Black and Laura J. Black, *Microbiology: Principles and Explorations*, 9th edition (Hoboken, NJ: John Wiley & Sons, 2015); and the List of Prokaryotic Names with Standing in Nomenclature (LPSN) website, bacterio.net. Nicholas Money's quote is from Nicholas P. Money, "Against the Naming of Fungi," *Fungal Biology* 117, no. 7–8 (2013): 463–65.

For information about doodlers, I used Jackie Andrade, "What Does Doodling Do?" *Applied Cognitive Psychology* 24, no. 1 (2010): 100–106, and I visited (along with over 8 million other people) "Epic Pen Spinning," YouTube video, 2:42, posted by Kuma Films, 2013, youtube.com/watch?v=voqYX8_VjvQ. Likewise, a search through YouTube for ASMR (autonomous sensory meridian response) will lead you to an enjoyable 15 minutes or so of listening to plastic crinkle.

Stephen Giovannoni's quote comes from "Many Challenges to Classifying Microbial Species," *Microbes and Evolution,* eds. Roberto Kolter and Stanley Maloy (Washington, DC: ASM Press, 2012). And here I am as a piece of pie: Newyorkmyc.org.

Chapter 4

For help determining my personality disorders, I went to MentalHelp.net and Amanda Green, "16 Totally Normal Phobias You Didn't Know Had Names," Mental Floss, February 1, 2016, mentalfloss.com/article/52126/16-totally-normal-phobias-youdidnt-know-had-names.

Mitochondrial disease is a profound ailment. I learned about it from the United Mitochondrial Disease Foundation at umdf.org. Eleonora Napoli et al., "Deficits in Bioenergetics and Impaired Immune Response in Granulocytes from Children with Autism," *Pediatrics* 133, no. 5 (2014): e1405–10, pediatrics.aappublications.org/content/early /2014/04/16/peds.2013-1545 illustrated mitochondrial deficits in children with autism.

I read more about Lynn Margulis and symbiogenesis than I reported on, but it is a great story, and hers was a great life: Dorion Sagan, ed., *Lynn Margulis: The Life and Legacy of a Scientific Rebel* (White River Junction, VT: Chelsea Green, 2012); Charles Mann, "Lynn Margulis: Science's Unruly Earth Mother," *Science* 252, no. 5004 (1991): 378–81; Francisco Carrapiço, "Can We Understand Evolution without Symbiogenesis?" in *Reticulate Evolution: Symbiogenesis, Lateral Gene Transfer, Hybridization and Infectious Heredity*, ed. Nathalie Gontier (New York: Springer, 2015); and, of course, her book *Symbiosis in Cell Evolution: Life and Its Environment on the Early Earth* (San Francisco: W. H. Freeman, 1981). For a little background on the brilliant but creepy groundbreaking biologist Constantin Merezhkowsky, I read Jan Sapp, Francisco Carrapiço, and Mikhail Zolotonosov, "Symbiogenesis: The Hidden Face of Constantin Merezh-kowsky," *History and Philosophy of the Life Sciences* 24 (2002): 413–40.

There are a number of hypotheses regarding the bacteria that became mitochon-dria. Lynn Margulis and Dorion Sagan's book *Microcosmos* explores some. So does John Archibald's *One Plus One Equals One*; Nick Lane's *The Vital Question*; Jan Sapp, *Evolution by Association: A History of Symbiosis* (New York: Oxford University Press, 1994); Ed Yong, *I Contain Multitudes: The Microbes within Us and a Grander View of Life* (New York: HarperCollins, 2016); Marcia Stone, "Oddly Microbial: Selfish Genes," *Small Things Considered* (blog), 2014, schaechter.asmblog.org/schaechter/2014/08/oddly-microbial-selfish-genes.html?utm_source=feedburner&utm_medium=feed&utm_campaign=Feed%3A+schaechter+%28Small+Things+Considered%29; and S. Marvin Friedman, "How an Endosymbiont Earns Tenure," *Small Things Considered* (blog), 2012, schaechter.asmblog.org/schaechter/2012/06/how-an-endosymbiont-earns-tenure.html. To learn more about how we inherit mitochondrial DNA, I referenced Steph Yin, "Why Do We Inherit Mitochondrial DNA Only from Our Mothers?," *New York Times*, June 23, 2016.

The nucleus is, as Moselio Schaechter told me, a huge problem. My biology book had its explanation, and I read others in T. Martin Embley and William Martin, "Eukaryotic Evolution, Changes and Challenges," *Nature* 440 (2006): 623–30; Carl Zim-mer, "On the Origin of Eukaryotes," *Science* 325, no. 5941 (2009): 666–68; and Tauana Junqueira Cunha, "Origins of Eukaryotes: Who Are Our Closest Relatives?," *Science in the News* (blog), 2014, sitn.hms.harvard.edu/flash/2014/origins-of-eukaryotes-who-are-our-closest-relatives. And new ideas keep coming, to wit, I learned about how the nucleus may have evolved to protect its DNA from pathogens from James F. White at Rutgers University.

Yes, onions have more DNA than you, according to the *Harvard Gazette*, February 10, 2000, news.harvard.edu/gazette/story/2000/02/why-onions-have-more-dna-than-you-do. Comparing genomes of different organisms is pretty humbling. I read more at genomenewsnetwork.org/articles/02_01/Sizing_genomes.shtml.

The tree-of-life model, where all life springs from a single trunk, has been debunked. I referred to "Ring of Life," *Astrobiology Magazine*, September 12, 2004, astrobio.net/extreme-life/ring-of-life; Carl Zimmer, "Scientists Unveil New 'Tree of Life,'" *New York Times*, April 11, 2016, nytimes.com/2016/04/12/science/scientists-unveil-new-tree-of-life.html; and James A. Lake, "Evidence for an Early Prokaryotic Endosymbiosis," *Nature* 460 (2009): 967–71. Symbiosis has been seen in action, as evi-denced by Kwang W. Jeon's incredible paper: "Bacterial Endosymbiosis in Amoebae," *Trends in Cell Biology* 5, no. 3 (1995): 137–40.

Chapter 5

Mnemonics has an interesting history, well described in Joshua Foer's *Moonwalk-ing with Einstein: The Art and Science of Remembering Everything* (New York: Penguin,

2012). There are a variety of websites that direct you to well-known mnemonics or that guide you through the different types of mnemonics, like acronyms, acrostics, rhymes, the keyword method, the image-name technique, chaining, and the method of loci. To learn more about memory, I consulted the USA Memory Championship website at usamemorychampionship.com, the Alzheimer's Association (alz.org), the National Institutes of Health's National Institute on Aging (nia.nih.gov/health), and the American Psychological Association (apa.org), as well as the short documentary, produced by Vivien Cardone and Jonathan Napolitano, *Ben Franklin Blowing Bubbles at a Sword: The Journeys of a Mental Athlete,* released April 2012. I cite some of the basics of brain science from thebrain.mcgill.ca, a website formerly financed by the Canadian Institutes of Health Research, accessed August 17, 2017, and Joseph Troncale, "Your Lizard Brain: The Limbic System and Brain Functioning," *Where Addiction Meets Your Brain* (blog), *Psychology Today,* April 22, 2014, psychologytoday.com/blog/where-addiction-meets-your-brain/201404/your-lizard-brain.

About multicellularity, I included information from Ed Yong, "Yeast Suggests Speedy Start for Multicellular Life," *Nature,* January 16, 2012, nature.com/news/yeast-suggests-speedy-start-for-multicellular-life-1.9810.

There are many excellent sources for the evolution of life in oceans. I used those I found most vivid: Lynn Margulis and Dorion Sagan's *Garden of Microbial Delights,* Richard Fortey's *Life,* and a visit to see the stromatolite fossils at the New York State Museum's Lester Park in Saratoga Springs (nysm.nysed.gov/research-collections /geology/resources/lester-park), as well as Aparna Vidyasagar, "What Are Protists?" Live Science, March 30, 2016, livescience.com/54242-protists.html. Facts about babesia were found on LymeDisease.org and in the Comparison Chart of Lyme Disease and Co-Infections Symptoms at lyme-symptoms.com/LymeCoinfectionChart.html. I learned about its discovery in Paul de Kruif's *Microbe Hunters.* The article by Aneri Pattani, "It's High Time for Ticks, Which Are Spreading Diseases Farther," *New York Times,* July 24, 2017, illustrates how prevalent babesiosis has become.

I found this study guide for protist classification from Mt. San Antonio College helpful: instruction2.mtsac.edu/mcooper/Biology%202/Labs/Protistalab1.pdf.

Slime mold intelligence facts were gleaned from "Slime Design Mimics Tokyo's Rail System: Efficient Methods of a Slime Mold Could Inform Human Engineers," ScienceDaily, January 22, 2010, accessed August 13, 2017, sciencedaily.com /releases/2010/01/100121141051.htm; "A Life of Slime: Network-Engineering Problems Can Be Solved by Surprisingly Simple Creatures," *Economist,* January 23, 2010, economist.com/node/15328524; Ferris Jabr, "How Brainless Slime Molds Redefine Intelligence," *Scientific American,* November 7, 2012, scientificamerican.com/article /brainless-slime-molds; Chris R. Reid et al., "Decision-Making without a Brain: How an Amoeboid Organism Solves the Two-Armed Bandit," *Journal of the Royal Society Interface* 13, no. 119 (2016): doi.org/10.1098/rsif.2016.0030; and the documentary film "The Creeping Garden," Cinema Iloobia, directed by Tim Grabham and Jasper Sharp, 2014, creepinggarden.com. For details about the amoeba that eats your brain, I referred to "Parasites—*Naegleria fowleri*—Primary Amebic Meningoencephalitis (PAM)—Amebic Encephalitis," Centers for Disease Control and Prevention, accessed August 17, 2017, cdc.gov/parasites/naegleria/index.html.

Information about the role of cortisol in stress and its correlation with the gut microbiome was gleaned from the following sources. For the effects of chronic stress, I learned from the Mayo Clinic at mayoclinic.org/healthy-lifestyle/stress-management/in-depth/stress/art-20046037; Ben Bernstein, "How Stress Affects Your Test Scores," *The College Puzzle* (blog), 2012, collegepuzzle.stanford.edu/?p=2242; Amy F. T. Arnsten, "Stress Signalling Pathways That Impair Prefrontal Cortex Structure and Function," *Nature Reviews Neuroscience* 10, no. 6 (2009): 410–22; and X. Zhao, R. L. Selman, and H. Haste, "Academic Stress in Chinese Schools and a Proposed Preventive Intervention Program," *Cogent Education* 2, no. 1 (2015). For the connection between the gut microbiome and production of stress hormones, I read L. Galland, "The Gut Microbiome and the Brain," *Journal of Medicinal Food* 17, no. 12 (2014): 1261–72; M. Carabotti et al., "The Gut-Brain Axis: Interactions between Enteric Microbiota, Central and Enteric Nervous Systems," *Annals of Gastroenterology* 28, no. 2 (2015), 203–209; and Andrea Anderson, "Why Do I Always Get Sick after Final Exams?," *Scienceline*, January 8, 2007, scienceline.org/2007/01/ask-anderson-finalscough.

The Weill Cornell study on bacteria in the NYC subway that caused all the hoopla is E. Afshinnekoo et al., "Geospatial Resolution of Human and Bacterial Diversity with City-Scale Metagenomics," *Cell Systems* 1, no. 1 (2015): 72–87. Corrections of some of the specifics of that study can be found in "Erratum," *Cell Systems* 1, no. 1 (2015): 97.

The paragraphs on genetic material exchange were gleaned from Matt Ridley, *The Red Queen: Sex and the Evolution of Human Nature* (New York: Harper Perennial, 1993); "Populations Survive Despite Many Deleterious Mutations: Evolutionary Model of Muller's Ratchet Explored," ScienceDaily, August 10, 2012, accessed August 17, 2017, sciencedaily.com/releases/2012/08/120810083613.htm; Megan Scudellari, "The Sex Paradox," *Scientist*, July 1, 2014, the-scientist.com/?articles.view/articleNo/40333/title/The-Sex-Paradox/; Elio Schaechter, "Sex and the Single Pillbug," *Small Things Considered* (blog), 2017, schaechter.asmblog.org/schaechter/2017/02/sex-and-the-single-pillbug.html?utm_source=feedburner&utm_medium=feed&utm_campaign=Feed%3A+schaechter+%28Small+Things+Considered%29; and L. Van Valen, "A New Evolutionary Law," *Evolution Theory* 1 (1973): 1–30, though I actually had to access this paper through Scribd at scribd.com/document/331418527/Vol-1-No-1-1-30-L-Van-Valen-A-new-evolutionary-law-pdf.

To describe the Burgess Shale, I read the tutorial on the Royal Ontario Museum's website: burgess-shale.rom.on.ca/en/index.php. The paper describing the oldest footprints on land is R. B. MacNaughton et al., "First Steps on Land: Arthropod Trackways in Cambrian-Ordovician Eolian Sandstone, Southeastern Ontario, Canada," *Geology* 30, no. 5 (2002): 391–94. I used *Life* by Richard Fortey and *The Amoeba in the Room* by Nicholas P. Money to flesh out my narrative of how life made landfall.

Chapter 6

This chapter was built from interviews with Timothy Crews, director of research and lead scientist at the Land Institute; James F. White, mycologist and plant pathologist at Rutgers University; Kristine Nichols, chief scientist, the Rodale Institute; and Michael Wood, proprietor of MykoWeb: Mushrooms, Fungi, Mycology (mykoweb .com). I also used the following books: *The Amoeba in the Room* by Nicholas P. Money, *Teaming with Microbes* by Jeff Lowenfels and Wayne Lewis, *The Hidden Half of Nature* by David R. Montgomery and Anne Biklé, *Dirt: The Ecstatic Skin of the Earth* by William Bryant Logan (New York: Riverhead Books, 1995), and *Soil Microbiology, Ecology, and Biochemistry,* 4th edition, by Eldor A. Paul (Cambridge, MA: Academic Press, 2014), as well as my own work in *Mycophilia.* Specific papers and websites follow.

The grade-grubbing site I reference is chronicle.com/forums/index.php /topic,31970.0.html.

The websites for the TerraGenome and Earth Microbiome projects are terragenome.org and earthmicrobiome.org, respectively. The quote about the numbers of microbes in the soil is from *Trees, Truffles, and Beasts: How Forests Function* by Chris Maser, Andrew W. Claridge, and James M. Trappe (New Brunswick, NJ: Rutgers University Press, 2008). Fungal diversity numbers change, but I used David L. Hawksworth and Robert Lücking, "Fungal Diversity Revisited: 2.2 to 3.8 Million Species," *Microbiology Spectrum* 5, no. 4 (2017): doi.org/10.1128/microbiolspec.FUNK-0052-2016. For numbers of plant species, I used the report *State of the World's Plants, 2016,* the Royal Botanic Gardens, Kew, stateoftheworldsplants.com/2016/report/sotwp_2016.pdf. Facts about mycorrhizae colonization are thanks to Moselio Schaechter, "Fine Reading: Unearthing the Roots of Ectomycorrhizal Symbioses," *Small Things Considered* (blog), 2017, schaechter.asmblog.org/schaechter/2017/04/fine-reading-unearthing-the-roots -of-ectomycorrhizal-symbioses.html?utm_source=feedburner&utm_medium=feed&utm _campaign=Feed%3A+schaechter+%28Small+Things+Considered%29. Michael Wood updated me on facts about fungal species numbers, and I quote Gary Lincoff from my book *Mycophilia.* I received insight into the rock-degrading capacities of fungi from Li Zibo et al., "Cellular Dissolution at Hypha- and Spore-Mineral Interfaces Revealing Unrecognized Mechanisms and Scales of Fungal Weathering," *Geology* 44, no. 4 (2016): 319–22 and Eric Hand, "Iron-Eating Fungus Disintegrates Rocks with Acid and Cellular Knives," *Science,* March 14, 2016, sciencemag.org/news/2016/03/iron-eating -fungus-disintegrates-rocks-acid-and-cellular-knives, and the sugar cube analogy comes from *Teaming with Microbes.* I learned about *Tortotubus* from Jacqueline Ronson, "This 'Tortotubus' Fungus Is the Oldest Land Fossil Ever Found," Inverse Science, March 4, 2016, inverse.com/article/12410-this-tortotubus-fungus-is-the-oldest-land -fossil-ever-found and Martin R. Smith, "Cord-Forming Paleozoic Fungi in Terrestrial Assemblages," *Botanical Journal* 180, no. 4 (2016): 452–60, and about the diversification of fungi from Francis Martin et al., "Unearthing the Roots of Ectomycorrhizal Symbioses," *Nature Reviews Microbiology* 14 (2016): 760–73.

I drew some information about lichens from Toby Spribille et al., "Basidiomycete Yeasts in the Cortex of Ascomycete Macrolichens," *Science,* July 21, 2016, doi.org /10.1126/science.aaf8287; Scott T. Bates et al., "Bacterial Communities Associated with the Lichen Symbiosis," *Applied and Environmental Microbiology* 77, no. 4 (2011) 1309–14; and "Lichen Can Survive in Space: Space Station Research Sheds Light on Origin Of Life; Potential for Better Sunscreens," ScienceDaily, June 23, 2012, accessed August 13, 2017, sciencedaily.com/releases/2012/06/120623145623.htm.

For the paragraphs about ancient terrestrial plants, I referenced Terry Devitt, "Ancestors of Land Plants Make the Leap to Shore," University of Wisconsin-Madison News, October 5, 2015, news.wisc.edu/ancestors-of-land-plants-were-wired-to-make -the-leap-to-shore and Katie J. Field et al., "Functional Analysis of Liverworts in Dual Symbiosis with Glomeromycota and Mucoromycotina Fungi under a Simulated Palaeozoic CO_2 Decline," *ISME Journal* 10, no. 6 (2016): 1514–26.

Lynn Margulis and Dorion Sagan's wonderful analogy of a seed being like an embryo waiting for a bullish economy is from their book *Microcosmos.*

Beyond the sources mentioned above, I used John Ingraham's *March of the Microbes* to describe nitrogen fixing and defixing, and to understand what happens to plants with insufficient nitrogen, I used "Guide to Symptoms of Plant Nutrient Deficiencies," University of Arizona Cooperative Extension, May 1999, extension.arizona .edu/sites/extension.arizona.edu/files/pubs/az1106.pdf.

It takes 1,000 years for the first centimeter of soil to form rock according to *Dirt* by William Bryant Logan.

I built my paragraph on glomalin from "Glomalin: Hiding Place for a Third of the World's Stored Soil Carbon," *Agricultural Research,* September 2002; Christina Kaiser, "Lazy Microbes Are Key for Soil Carbon and Nitrogen Sequestration," International Institute for Applied Systems Analysis, December 1, 2015, iiasa.ac.at/web/home/about /news/151201-microbe-soil.html; Mark Ridley's book *The Cooperative Gene: How Mendel's Demon Explains the Evolution of Complex Beings* (New York: The Free Press, 2001); and *Teaming with Microbes* by Jeff Lowenfels and Wayne Lewis. Facts about carbon sequestration in soil came from C. Le Quéré et al., "Global Carbon Budget 2013," *Earth System Science Data* 6 (2014): 235–63, doi.org/10.5194/essd-6-235-2014.

The leaching of plant sugars into soil communities was well described in Soil Science Society of America's "Soil Biology" at soils4teachers.org/biology-life-soil; Marie Lescroart, "Exploring the Soil's Genetic Biodiversity," *CNRS International Magazine,* accessed August 13, 2017, www2.cnrs.fr/en/1600.htm; the special Focus issue on Plant–Microbe Interactions, *Nature Reviews Microbiology* 11, no. 799 (2013); "Scientists Discover That Symbiotic Fungi Get Carbon from Plants in the Form of Fatty Acids," *Crop Biotech Update,* June 14, 2017, isaaa.org/kc/cropbiotechupdate/article/default .asp?ID=15513; and Mike Amaranthus and Bruce Allyn, "Healthy Soil Microbes, Healthy People: The Microbial Community in the Ground Is as Important as the One in Our Guts," *Atlantic,* June 11, 2013, theatlantic.com/health/archive/2013/06/healthy -soil-microbes-healthy-people/276710.

To report on the fungal–bacterial symbiosis, I read P. Frey-Klett, J. Garbaye, and M. Tarkka, "The Mycorrhiza Helper Bacteria Revisited," *New Phytologist* 176 (2007): 22–36 and Jacinta Gahan and Achim Schmalenberger, "The Role of Bacteria and Mycorrhiza in Plant Sulfur Supply," *Frontiers in Plant Science* 5 (2014): 723.

Further, I referred to Greg Reid and Percy Wong, "Soil Bacteria," *Soil Biology Basics*, accessed August 13, 2017, dpi.nsw.gov.au/__data/assets/pdf_file/0017/41642 /Soil_bacteria.pdf. I paraphrased the descriptions of the soil microbiome in Matt Soniak, "The Tiniest Capitalists: How Microbes Replicate the Market," *Week*, January 24, 2014, theweek.com/articles/452338/tiniest-capitalists-how-microbes-replicate-market. The examples I used showing how bacteria and fungi help each other were from P. Frey-Klett et al., "Bacterial-Fungal Interactions: Hyphens between Agricultural, Clinical, Environmental, and Food Microbiologists," *Microbiology and Molecular Biology Reviews* 75, no. 4 (2011): 583–609.

Facts about soil protists were found in Arlene J. Tugel, Ann Lewandowski, and Deb Happe-vonArb, eds., *Soil Biology Primer*, revised edition (Ankeny, IA: Soil and Water Conservation Society, 2000) and facts about nematodes were found in *Teaming with Microbes* by Jeff Lowenfels and Wayne Lewis and "Featured Creatures," University of Florida, Institute of Food and Agricultural Sciences, accessed August 13, 2017, entnemdept.ufl.edu/creatures/nematode/soil_nematode.htm.

I found Nathan Cobb's quote in Laetitia Gunton, "A Worm's World," *Biologist* 64, no. 3 (2017): 18–21, thebiologist.rsb.org.uk/biologist/158-biologist/features/1762-this -is-a-worm-s-world, along with other facts about nematodes.

The tardigrade footnote was written from a few sources: Nicholas Bakalar, "Tardigrades Have the Right Stuff to Resist Radiation," *New York Times*, September 26, 2016, and Sara Sánchez-Moreno, Howard Ferris, and Noemí Guil, "Role of Tardigrades in the Suppressive Service of a Soil Food Web," *Agriculture, Ecosystems & Environment* 124, no. 3–4 (2008): 187–92.

The earthworm paragraph was constructed with help from Amy Stewart, *The Earth Moved: On the Remarkable Achievements of Earthworms* (New York: Workman, 2012).

I used many sources to get the ADD story straight: Ryan D'Agostino, "The Drugging of the American Boy," *Esquire*, April 2014; A. D. DeSantis, E. M. Webb, and S. M. Noar, "Illicit Use of Prescription ADHD Medications on a College Campus: A Multimethodological Approach," *Journal of American College Health* 57, no. 3 (2008): 315–24; Henry Greely et al., "Towards Responsible Use of Cognitive-Enhancing Drugs by the Healthy," *Nature* 456 (2008): 702–705; Kimberly Holland and Elsbeth Riley, "ADHD by the Numbers: Facts, Statistics, and You," Healthline, September 4, 2014, healthline.com/health/adhd/facts-statistics-infographic#1; William C. Watkins, "Prescription Drug Misuse among College Students: A Comparison of Motivational Typologies," *Journal of Drug Issues* 46, no. 3 (2016): doi.org/10.1177/0022042616632268; "Focalin vs Adderall—A Side-by-Side Comparison," Brain Pro Tips, October 1, 2016, accessed August 13, 2017, brainprotips.com/focalin-vs-adderall; and Alan Schwarz,

ADHD Nation: Children, Doctors, Big Pharma, and the Making of an American Epidemic (New York: Scribner, 2016).

I watched a Tibetan sky burial: "Traditional Tibetan Sky Burial," YouTube video, 10:54, posted by edilcanidal 2016, youtube.com/watch?v=LhixERFx9PA. I'm not the only one who thinks the Infinity Burial Suit is unlikely to speed decomposition: Sarah Crews and Tim Crews, "Eco-Friendly 'Burial Suits' Are on the Rise, but Are They Really Necessary?," AlterNet, March 20, 2016, alternet.org/environment/eco-friendly-burial -suits-are-rise-are-they-really-necessary. Regarding decomposition, *Dirt* by William Bryant Logan provided a colorful description. I also used the following: Jessica L. Metcalf et al., "A Microbial Clock Provides an Accurate Estimate of the Postmortem Interval in a Mouse Model System," *eLife* 2 (2013): e01104, doi.org/10.7554/eLife.01104 and R. A. Benner Jr., W. F. Staruszkiewicz, and W. S. Otwell, "Putrescine, Cadaverine, and Indole Production by Bacteria Isolated from Wild and Aquacultured Penaeid Shrimp Stored at 0, 12, 24, and 36 Degrees C," *Journal of Food Protection* 67, no. 1 (2004): 124–33. Jennifer DeBruyn was quoted from Rachel Ehrenberg, "Microbes May Be a Forensic Tool for Time of Death," *Science News,* July 22, 2015, sciencenews.org/blog /culture-beaker/microbes-may-be-forensic-tool-time-death. The connection between soil and the etymology of biblical names was made by the Israeli soil scientist Daniel Hillel, *Out of the Earth: Civilization and the Life of the Soil* (Berkeley: University of California Press, 1992).

Chapter 7

I wrote this chapter based on interviews with Jim Gillespie, director, Division of Soil Conservation & Water Quality, Iowa Department of Agriculture; David Ertl from the Iowa Corn Promotion Board; Don Smith from Kiss the Ground; farmer Ron Heck; Michael, Roni, and Jennifer Sweeney, Krista McGuire, Barnard College, Columbia University; Timothy Crews of the Land Institute; and James F. White of Rutgers University.

The garage lab movement was reported on by Jed Lipinski, "Turning Geek into Chic," *New York Times,* December 17, 2010, who also reported on the building guru Al Attara in "On Flatbush Avenue, Seven Stories Full of Ideas," *New York Times,* January 11, 2011. Ellen Jorgensen's TED talk "Biohacking—You Can Do It, Too" is posted at ted .com/talks/ellen_jorgensen_biohacking_you_can_do_it_too. I found American college dropout statistics at collegeatlas.org/college-dropout.html and learned more about the phenomenon from Erin Dunlop Velez, *America's College Drop-Out Epidemic: Understanding the College Drop-Out Population*, working paper 109, National Center for Analysis of Longitudinal Data in Education Research, January 2014, air.org/sites /default/files/downloads/report/AIR-CALDER-Understanding-the-College-Dropout -Population-Jan14.pdf and "Freshman Retention Rate: National Universities," *U.S. News & World Report,* accessed August 13, 2017, usnews.com/best-colleges/rankings /national-universities/freshmen-least-most-likely-return.

The paragraphs on soil erosion were built from information found in Dan Barber,

NOTES

241NOTES 241

The Third Plate: Field Notes on the Future of Food (New York: Penguin, 2014); Vernon Gill Carter and Tom Dale, *Topsoil and Civilization* (Norman: University of Oklahoma Press, 1974); and Timothy Egan, *The Worst Hard Time: The Untold Story of Those Who Survived the Great American Dust Bowl* (Boston: Mariner, 2006). I also used lectures by Wes Jackson, cofounder of the Land Institute; Peter Kenmore, former agricultural entomologist at the Food and Agriculture Organization of the United Nations; Angus Wright, cofounder of Environmental Studies at California State University, Sacramento; and Ricardo Salvador from the Union of Concerned Scientists at the Land Institute Prairie Festival, 2015. Papers that were useful were: "Erosion's Long, Destructive Train," Environmental Working Group, accessed August 13, 2017, ewg.org/losingground /report/erosions-long-destructive-train.html; H. Eswaran, R. Lal, and P. F. Reich, "Land Degradation: An Overview," USDA Natural Resources Conservation Service Soils, accessed August 13, 2017, nrcs.usda.gov/wps/portal/nrcs/detail/soils/use/?cid=nrcs142p2 _054028; David R. Montgomery, "Soil Erosion and Agricultural Sustainability," *PNAS* 104, no. 33 (2007): 13268–72; Tom Philpott, "Iowa Is Getting Sucked into Scary Vanishing Gullies," *Mother Jones*, February 7, 2014, motherjones.com/food/2014/02/iowas -vaunted-farms-are-losing-topsoil-alarming-rate; and Charles C. Mann, *1493: Uncovering the New World Columbus Created* (New York: Knopf, 2011). Regarding Jethro Tull and tillage, I used "Tillage Management and Soil Organic Matter," Iowa State University, 2005, publications.iowa.gov/2811/ and Jethro Tull, *Horse-Hoeing Husbandry* (London: William Cobbett, 1829), eminencegris.com/TULL/Tull.htm.

The harm fertilizer does to soil microbes comes from the following sources: *The Hidden Half of Nature* by David R. Montgomery and Anne Biklé; "Cover Crops to Improve Soil in Prevented Planting Fields," USDA Natural Resources Conservation Service, accessed August 13, 2017, https://www.nrcs.usda.gov/Internet/FSE _DOCUMENTS/stelprdb1142408.pdf; "Does Fertilizer Harm Soil Microbes?," *Agri-Briefs*, no. 1 (Spring 2004): www.ipni.net/ppiweb/agbrief.nsf/5a4b8be72a35cd468525 68d9001a18da/90d25af635efb8d685256e4a0072e3de!OpenDocument; and John Sawyer and John Lundvall, "Nitrogen Use in Iowa Corn Production," Iowa State University Extension and Outreach, CROP 3073, April 2007, store.extension.iastate.edu /Product/14281. The controversy of Haber and Bosch's Nobel is argued in ScienceHeroes.com, scienceheroes.com/index.php?option=com_content&view =article&id=28&Itemid=58, and Tom Philpott, "A Brief History of Our Deadly Addiction to Nitrogen Fertilizer," *Mother Jones*, April 19, 2013, motherjones.com /food/2013/04/history-nitrogen-fertilizer-ammonium-nitrate.

Facts about dead zones were found at Robert J. Diaz and Rutger Rosenberg, "Spreading Dead Zones and Consequences for Marine Ecosystems," *Science* 321, no. 5891 (2008): 926–29 and "Mississippi River/Gulf of Mexico Hypoxia Task Force," US Environmental Protection Agency, accessed August 13, 2017, epa.gov/ms-htf.

Facts in the paragraph on the disadvantages of using herbicides were found in A. Sebiomo, V. W. Ogundero, and S. A. Bankole, "Effect of Four Herbicides on Microbial Population, Soil Organic Matter and Dehydrogenase Activity," *African Journal of*

Biotechnology 10, no. 5 (2011): 770–78 and Stephanie Strom, "Misgivings about How a Weed Killer Affects the Soil," *New York Times*, September 19, 2013.

About postcollege careers, I used the following sources: Gwen Moran, "How You May Be Unconsciously Shaping Your Child's Career Choices," Dedman College of Humanities and Sciences blog, 2015, blog.smu.edu/dedmancollege/2015/06/22/george-holden-psychology-2010-study-cited-in-a-story-about-parents-unconscious-influences-on-childs-career-interests; Brad Waters, "The Immense Pressure of Career Choice," *Design Your Path* (blog), *Psychology Today*, 2012, psychologytoday.com/blog/design-your-path/201206/the-immense-pressure-career-choice; Linda Olshina Lavine, "Parental Power as a Potential Influence on Girls' Career Choice," *Child Development* 53, no. 3 (1982): 658–63; "Maximizing Millennials in the Workplace," *UNC Executive Development Blog*, 2014, execdev.kenan-flagler.unc.edu/blog/maximizing-millennials-in-the-workplace; Karen K. Meyers and Kamyab Sadaghiani, "Millennials in the Workplace: A Communication Perspective on Millennials' Organizational Relationships and Performance," *Journal of Business and Psychology* 25, no. 2 (2010): 225–38; and "Public Transit Is a Deciding Factor for Millennials in the Workforce," press release, *Mother Earth News*, June 3, 2014, accessed August 13, 2017, motherearthnews.com/green-transportation/millennials-in-the-workforce-zb0z1405zwea.

For the paragraphs on the organic movements, I used, among others, the following resources: Matthew Reed, *Rebels for the Soil: The Rise of the Global Organic Food and Farming Movement* (London: Earthscan, 2010); Charles C. Mann, "Our Good Earth," *National Geographic*, September 2008; "Terra Preta de Indio," Cornell University, Department of Crop and Soil Sciences, accessed August 13, 2017, css.cornell.edu/faculty/lehmann/research/terra%20preta/terrapretamain.html; James E. McWilliams, "All Sizzle and No Steak: Why Allan Savory's TED Talk about How Cattle Can Reverse Global Warming Is Dead Wrong," *Slate*, April 22, 2013, slate.com/articles/life/food/2013/04/allan_savory_s_ted_talk_is_wrong_and_the_benefits_of_holistic_grazing_have.html; Mark Keating, "Interview: Grace Gershuny on the Past, Future of Organics," *Acres U.S.A.*, July 2016; Grace Gershuny, *Organic Revolutionary: A Memoir of the Movement for Real Food, Planetary Healing, and Human Liberation* (Joes Brook Press, 2016); Lady Eve Balfour, "Towards a Sustainable Agriculture—The Living Soil," soilandhealth.org/wp-content/uploads/01aglibrary/010116Balfourspeech.html; E. B. Balfour, *The Living Soil and the Haughley Experiment* (Basingstoke, England: Palgrave Macmillan, 1976); "Who Was Rudolf Steiner?," Biodynamic Association, biodynamics.com/steiner.html; Rudolf Steiner, *Agriculture: An Introductory Reader* (Forest Row, England: Sophia Books, 2003); and an interview with Lance Hanson of Jack Rabbit Hill Winery in Hotchkiss, Colorado.

Information about using microbes in agriculture came from a variety of sources, including Richard Conniff, "Super Dirt," *Scientific American*, September 2013; Carl Zimmer, "Scientists Hope to Cultivate an Immune System for Crops," *New York Times*, June 16, 2016; Geoffrey Mohan, "Will Microbes Save Agriculture?," *Los Angeles Times*, May 28, 2016, latimes.com/business/la-fi-soil-microbes-20160527-snap-story.html;

Trade and Environment Review 2013: Wake Up Before It Is Too Late: Make Agriculture Truly Sustainable Now for Food Security in a Changing Climate, United Nations Conference on Trade and Development, unctad.org/en/PublicationsLibrary/ditcted2012d3 _en.pdf; and Timothy E. Crews et al., "Going Where No Grains Have Gone Before: From Early to Mid-Succession," *Agriculture, Ecosystems & Environment* 223 (2016): 223–38. I learned about how farms might mitigate erosion and hypoxia from Verena Seufert and Navin Ramankutty, "Many Shades of Gray—The Context-Dependent Performance of Organic Agriculture," *Science Advances* 3, no. 3 (2017): doi.org/10.1126/sciadv.1602638.

Chapter 8

For my reporting on the tutor business in New York City, I used the following resources: Caroline Moss, "Meet the Guy Who Makes $1,000 an Hour Tutoring Kids of Fortune 500 CEOs over Skype," Business Insider, August 26, 2014, businessinsider.com /anthony-green-tutoring-2014-8; David Ludwig, "Are Tutors the New Waiters?," *Atlantic*, January 27, 2015, theatlantic.com/education/archive/2015/01/are-tutors-the-new -waiters/384745/; and Jenny Anderson, "Push for A's at Private Schools Is Keeping Costly Tutors Busy," *New York Times*, June 7, 2011. For more Columbia student rhymes about bathrooms, bwog.com/2014/09/12/where-are-all-the-bathrooms/. Barack Obama made the comment about "acting white" in his keynote address at the Democratic National Convention, 2004 (seems so long ago). Natalie Boelman posted her reports in *Scientist at Work: Notes from the Field* (blog), *New York Times*, 2011, scientistatwork .blogs.nytimes.com/author/natalie-boelman/. Nicholas P. Money's wonderful quote is in his book *The Amoeba in the Room*. The footnote about bees is from Lawrence Goodwyn, "Flowers Serve as Hubs for 'Friendly' Bacteria," *Daily Texan*, September 22, 2016, dailytexanonline.com/2016/09/22/flowers-serve-as-hubs-for-%E2%80%98friendly%E 2%80%99-bacteria.

The facts in the paragraphs on the microbiome of plants are drawn from many sources. Among them, *The Hidden Half of Nature* by David R. Montgomery and Anne Biklé; EnDoBiodiversity.org, which focuses on endophytic fungi in boreal forests; my own *Mycophilia*; Rakesh Santhanam et al., "Native Root-Associated Bacteria Rescue a Plant from a Sudden-Wilt Disease That Emerged during Continuous Cropping," *PNAS* 112, 36 (2015): E5013–20; Stéphane Hacquard et al., "Microbiota and Host Nutrition across Plant and Animal Kingdoms," *Cell Host & Microbe* 17, no. 5 (2015): 603–16; David J. Newman and Gordon M. Cragg, "Endophytic and Epiphytic Microbes as 'Sources' of Bioactive Agents," *Frontiers in Chemistry* 3 (2015): 34; Bernard R. Glick, *Beneficial Plant-Bacterial Interactions* (New York: Springer, 2015); Maren L. Friesen et al., "Microbially Mediated Plant Functional Traits," *Annual Review of Ecology, Evolution, and Systematics* 42 (2011): 23–46; Gabriele Berg et al., "Unraveling the Plant Microbiome: Looking Back and Future Perspectives," *Frontiers in Microbiology* 5 (2014): 148, doi.org/10.3389/fmicb.2014.00148; F. V. Nunes and I. S. de Melo, "Isolation and Characterization of Endophytic Bacteria of Coffee Plants and Their Potential in

Caffeine Degradation," in *Environmental Toxicology*, eds. A. Kungolos et al. (WIT Press, 2006); Thomas R. Turner, Euan K. James, and Philip S. Poole, "The Plant Microbiome," *Genome Biology* 14, no. 6 (2013): 209; and Marcel G. A van der Heijden and Martin Hartmann, "Networking in the Plant Microbiome," *PLoS Biology* 14, no. 2 (2016): e1002378, doi.org/10.1371/journal.pbio.1002378.

To write the paragraphs about the role microbes play in plant defense and nutrition, I used the following, among others: X. Li et al., "The Endophytic Bacteria Isolated from Elephant Grass (*Pennisetum purpureum* Schumach) Promote Plant Growth and Enhance Salt Tolerance of Hybrid Pennisetum," *Biotechnology for Biofuels* 9, no. 1 (2016): 190; Chris Samoray, "Plants Trick Bacteria into Attacking Too Soon," *Science News*, February 9, 2016, sciencenews.org/article/plants-trick-bacteria-attacking-too -soon; Ed Yong, "Plants Use Fungi to Eat Insects by Proxy," *Not Exactly Rocket Science* (blog), *Discover*, 2012, blogs.discovermagazine.com/notrocketscience/2012/06/21 /plants-use-fungi-to-eat-insects-by-proxy/#.WYerxlKZNzg; Chanyarat Paungfoo-Lonhienne et al., "Turning the Table: Plants Consume Microbes as a Source of Nutrients," *PLoS ONE* 5, no. 7 (2010): e11915, doi.org/10.1371/journal.pone.0011915; D. Johnston-Monje and M. N. Raizada, "Plant and Endophyte Relationships: Nutrient Management," in *Comprehensive Biotechnology*, 2nd edition, volume 4, ed. Murray Moo-Young (Oxford, England: Pergamon, 2011); Mary L. Martialay, "Bacteria Make Natural Pigment from Simple Sugar," RPI News, June 29, 2017, accessed August 13, 2017, news.rpi.edu/content/2017/06/29/bacteria-make-natural-pigment-simple-sugar; Jutta Ludwig-Müller, "Plants and Endophytes: Equal Partners in Secondary Metabolite Production?," *Biotechnology Letters* 37, no. 7 (2015): 1325–34; Barbara Siegmund et al., "The Flavour of Strawberries: Can We Enhance It in a Natural Way?," poster presentation, Graz University of Technology, snoe.boku.ac.at/wp-content/uploads/2010/02 /Strawberry-flavour-Siegmund-et-al-Pangborn-Symposium-2009.pdf; and Susan Langthorp, "*Science*Shot: Fungi Provide an Early Warning System for Plants," *Science*, May 17, 2013, sciencemag.org/news/2013/05/scienceshot-fungi-provide-early-warning -system-plants. For a definition of oxidative stress, I used D. J. Betteridge, "What Is Oxidative Stress?," *Metabolism* 49, no. 2, supplement 1 (2000): 3–8.

My paragraphs on fungal endophytes were drawn from interviews with Russell Rodriguez and Regina Redman of Adaptive Symbiotic Technologies, James F. White of Rutgers University, and Rutgers PhD student Camille English; K. Saikkonen et al., "Fungal Endophytes: A Continuum of Interactions with Host Plants," *Annual Review of Ecology and Systematics* 29 (1998): 319–43; Gary Strobel and Bryn Daisy, "Bioprospecting for Microbial Endophytes and Their Natural Products," *Microbiology and Molecular Biology Reviews* 67, no. 4 (2003): 491–502; R. J. Rodriguez et al., "Fungal Endophytes: Diversity and Functional Roles," *New Phytologist* 182, no. 2 (2009): 314–30; A. Elizabeth Arnold, "Hidden within Our Botanical Richness, a Treasure Trove of Fungal Endophytes," *The Plant Press*, November 2008, cals.arizona.edu/mycoherb /arnoldlab/Arnold.PlantPress.pdf; Niina Heikkinen, "Fungus May Save Crops From Disease and Global Warming," *Scientific American,* February 17, 2015, scientificamerican

.com/article/fungus-may-save-crops-from-disease-and-global-warming/; P. Vanden-koornhuyse et al., "The Importance of the Microbiome of the Plant Holobiont," *New Phytologist* 206, no. 4 (2015): 1196–206; R. Rodriguez et al., "Stress Tolerance in Plants via Habitat-Adapted Symbiosis," *ISME Journal* 2, no. 4 (2008): 404–16; R. Rodriguez et al., "Symbiotic Regulation of Plant Growth, Development and Reproduction," *Communicative & Integrative Biology* 2, no. 2 (2009): 141–43; R. Rodriguez and R. Redman, "More Than 400 Million Years of Evolution and Some Plants Still Can't Make It on Their Own: Plant Stress Tolerance via Fungal Symbiosis," *Journal of Experimental Botany* 59, no. 5 (2008): 1109–14; Peter R. Atsatt and Matthew D. Whiteside, "Novel Symbiotic Protoplasts Formed by Endophytic Fungi Explain Their Hidden Existence, Lifestyle Switching, and Diversity within the Plant Kingdom," *PLoS ONE* 9, no. 4 (2014): e95266, doi.org/10.1371/journal.pone.0095266; Scott W. Behie and Michael J. Bidochka, "Nutrient Transfer in Plant–Fungal Symbiosis," *Trends in Plant Science* 19, no. 11 (2014): 734–40; Dan Cossins, "Plants Communicate with Help of Fungi," *Scientist,* May 14, 2013, the-scientist.com/?articles.view/articleNo/35542/title/Plants-Communicate-with-Help-of-Fungi; and Marie Bourgiugnon et al., "Ecophysiological Responses of Tall Fescue Genotypes to Fungal Endophyte Infection, Elevated Temperature, and Precipitation," *Crop Science* 55, no. 6 (2015): 2895–909. The statement that plants without endophytic fungi have returned to being aquatic comes from *Microcosmos* by Lynn Margulis and Dorion Sagan. Brendan Mormile worked on the study about loline while at the Microbial Ecology Lab at Southern Connecticut State University. He explained the work to me in an interview.

I wrote the pages about corn in Iowa from a variety of sources. They include: Betty Fussell, *The Story of Corn* (Albuquerque: University of New Mexico Press, 2004); iowacorn.org; worldofcorn.com; David Widmar, "Plants per Acre: A Look at Corn Plant Population," *Agricultural Economic Insights*, April 11, 2016, ageconomists.com/2016/04/11/plants-per-acre-look-corn-plant-population-trends/; Kiersten Wise and Daren Mueller, "Are Fungicides No Longer Just for Fungi? An Analysis of Foliar Fungicide Use in Corn," *APSnet Features* (2011): doi.org/10.1094/APSnetFeature-2011-0531. About glyphosate, "Scientists Agree: Glyphosate Is a Pesticide of High Concern," The Endocrine Disruption Exchange (TEDX), endocrinedisruption.org/enews/2016/02/15/scientists-agree-glyphosate-is-a-pesticide-of-high-concern/; Robert Kremer, "GMOs, Glyphosate and Soil Biology," podcast 47:00, posted by Food Integrity Now, 2015, foodintegritynow.org/2015/04/15/dr-robert-kremer-gmos-glyphosate-and-soil-biology/; Anthony Samsel and Stephanie Seneff, "Glyphosate, Pathways to Modern Diseases II: Celiac Sprue and Gluten Intolerance," *Interdisciplinary Toxicology* 6, no. 4 (2013): 159–84; and M. Druille et al., "Glyphosate Vulnerability Explains Changes in Root-Symbionts Propagules Viability in Pampean Grasslands," *Agriculture, Ecosystems & Environment* 202 (2015): 48–55. The quote to "build on" Roundup comes from the company website enlist.com for Enlist traits (as in genetic traits) technology, accessed August 13, 2017, enlist.com/en/how-it-works/enlist-traits. The percentage of glyphosate and Bt modified corn grown in the USA comes from Monsanto: monsanto.com/app

/uploads/2017/06/Glyphosate-benefits-and-safety-FINAL-Gov-Officials-ONLY.pdf.

On GMO corn, Wanqing Zhou, "Genetically Modified Crop Industry Continues to Expand," Worldwatch Institute, March 24, 2015, vitalsigns.worldwatch.org/vs-trend /genetically-modified-crop-industry-continues-expand; N. L. Swanson et al., "Genetically Engineered Crops, Glyphosate, and the Deterioration of Health in the United States," *Journal of Organic Systems* 9, no. 2 (2014); "Genetically Modified Corn Affects Its Symbiotic Relationship with Non-Target Soil Organisms," Phys.org, April 17, 2012, phys.org/news/2012-04-genetically-corn-affects-symbiotic-relationship.html; and Sean B. Carroll, "Tracking the Ancestry of Corn Back 9,000 Years," *New York Times*, May 24, 2010. I read the story of the ongoing Corn-Oil Experiment in Graham Bell, *Selection: The Mechanism of Evolution,* 2nd edition (New York: Oxford University Press, 2008). To listen to corn grow, "Listen and Watch Corn Grow," YouTube video, 1:02, posted by UNL CropWatch, 2016, youtube.com/watch?v=76xEkEXI2a4. Bacterial DNA is in most organisms, according to Ed Yong, "Bacterial DNA in Human Genomes," *Scientist*, June 20, 2013, the-scientist.com/?articles.view/articleNo/36108/title/Bacterial-DNA-in -Human-Genomes/. Regarding genetics and nutrition in plant foods, I read Jo Robinson, "Breeding the Nutrition out of Our Food," *New York Times*, May 25, 2013; *The Hidden Half of Nature* by David R. Montgomery and Anne Biklé; C. Claiborne Ray, "A Decline in the Nutritional Value of Crops," *New York Times*, September 12, 2015; and Jonathan Eisen, "Microbes in Food and Agriculture," YouTube video, 2:00:39, posted by Science Says, 2015, youtube.com/watch?v=GIvl6QEjHAk. Wendell Berry's quote comes from his widely published essay, "The Pleasures of Eating."

The Sterling Education guideline on keeping your audience awake can be found at sterlingeducation.com/the-sterling-blog/bid/70545/10-Easy-Steps-to-Lull-an -Audience-to-Sleep.html. Information on reductions in biomass due to atmospheric warming was found in S. D. Frey et al., "Microbial Biomass, Functional Capacity, and Community Structure after 12 Years of Soil Warming," *Soil Biology & Biochemistry* 40, no. 11 (2008): 2904–907; Aspen T. Reese, "How Climate Change Endangers Microbes— and Why That's Not a Good Thing," *Guest Blog, Scientific American*, 2016, blogs.scientificamerican.com/guest-blog/how-climate-change-endangers-microbes -and-why-that-s-not-a-good-thing/; and Yuting Liang et al., "Long-Term Soil Transplant Simulating Climate Change with Latitude Significantly Alters Microbial Temporal Turnover," *ISME Journal* 9 (2015): 2561–72 (and there are many other studies on this subject).

In my paragraphs about the new biotech, I used the following sources: interviews with Russell Rodriguez and Regina Redman of Adaptive Symbiotic Technologies; Carl Zimmer, "Scientists Hope to Cultivate an Immune System for Crops," *New York Times*, June 16, 2016; Mark Martin, "All for One, and One for All!" *Small Things Considered* (blog), 2009, schaechter.asmblog.org/schaechter/2009/10/all-for-one-and-one-for-all .html; John Upton, "How a New Evolutionary Map Could Help Farmers Eliminate Fertilizer," *Pacific Standard*, June 13, 2014, psmag.com/environment/nitrogen-83397; Amy Coombs, "Fighting Microbes with Microbes," *Scientist*, January 1, 2013, the-scientist .com/?articles.view/articleNo/33703/title/Fighting-Microbes-with-Microbes/.

A series of interviews with James F. White from Rutgers University illuminates the paragraph on the seed as the future of the plant. On the microbial effect on flavors of food, I refer to Carolyn Beans, "Demystifying Terroir: Maybe It's the Microbes Making Magic in Your Wine," *The Salt*, June 17, 2016, npr.org/sections/thesalt/2016/06/17 /482315073/demystifying-terroir-maybe-its-the-microbes-making-magic-in-your-wine; an interview with mycologist Tom Volk; Mary Beth Albright, "Organic Foods Are Tastier and Healthier, Study Finds," *The Plate* (blog), July 24, 2014, theplate.nationalgeographic .com/2014/07/14/organic-foods-are-tastier-and-healthier-study-finds/; Chris Zimmerman, "Using Ancient Genes to Improve Modern Crops," American Society of Agronomy, January 16, 2012, agronomy.org/science-news/using-ancient-genes-improve -modern-crops; and Jonathan W. Leff and Noah Fierer, "Bacterial Communities Associated with the Surfaces of Fresh Fruits and Vegetables," *PLoS ONE* 8, no. 3 (2013): e59310, doi.org/10.1371/journal.pone.0059310. James F. White showed me photos of microbes entering and exiting the root cells of a tomato plant. It was perception-altering. Like seeing a ghost.

Chapter 9

To read the lyrics for Frank Zappa's "Valley Girl," go to azlyrics.com/lyrics /frankzappa/valleygirl.html, or better, get the album, *Ship Arriving Too Late to Save a Drowning Witch*, Barking Pumpkin Records, 1982. Uptalk references are from James Gorman, "On Language; Like, Uptalk?" *New York Times*, August 15, 1993; Matt Seaton, "Word Up," *Guardian*, September 21, 2001, theguardian.com/books/2001/sep/21 /referenceandlanguages.mattseaton; and "The Unstoppable March of the Upward Inflection?," *BBC Magazine*, August 11, 2014, bbc.com/news/magazine-28708526. The video we saw on Yellowstone Park was "How Wolves Change Rivers," YouTube video, 4:33, posted by Sustainable Human, 2014, youtube.com/watch?v=ysa5OBhXz-Q.

My paragraph on the cascading effects of pulling microbes out of the food system is supported by Cheryl Long, "Industrially Farmed Foods Have Lower Nutritional Content," *Mother Earth News*, June/July 2009, motherearthnews.com/nature-and -environment/nutritional-content-zmaz09jjzraw; Dilfuza Egamberdiyeva, "The Effect of Plant Growth Promoting Bacteria on Growth and Nutrient Uptake of Maize in Two Different Soils," *Applied Soil Ecology*, 36, no. 2–3 (2007): 184–89; and *The Hidden Half of Nature* by David R. Montgomery and Anne Biklé.

About cows, I used these sources among others: Betty Fussell, *Raising Steaks: The Life and Times of American Beef* (Orlando: Harcourt, 2008); Nicolette Hahn Niman, *Defending Beef: The Case for Sustainable Meat Production* (White River Junction, VT: Chelsea Green Publishing, 2014); R. J. Wallace, "Ruminal Microbiology, Biotechnology, and Ruminant Nutrition: Progress and Problems," *Journal of Animal Science* 72, no. 11 (1994): 2992–3003; and "Factory Farm Nation: 2015 Edition," Food & Water Watch, May 27, 2015, accessed August 13, 2017, foodandwaterwatch.org/insight/factory-farm -nation-2015-edition. I visited "Ask a Farmer" to answer questions like "How are antibiotics used in cattle?" "Are feedlot cattle fed antibiotics and hormones?" and "Does feeding corn harm cattle?" at agricultureproud.com. Regarding methane emissions, I referred to "Carbon, Methane Emissions and the Dairy Cow," PennState Extension, August 14, 2017, extension.psu.edu/animals/dairy/nutrition/nutrition-and-feeding /diet-formulation-and-evaluation/carbon-methane-emissions-and-the-dairy-cow and "Case Study—Science Proves It's Possible to Breed Green Livestock," Australian Government, Department of Agriculture and Water Resources, 2015, agriculture.gov.au /ag-farm-food/climatechange/climate/communication/factsheets-case-studies-and-dvds /case_studies/case-study-science-proves-it-as-possible-to-breed-green-livestock.

Butler library stories were gleaned from Ben Ratliff, "Grazing in the Stacks of Academe," *New York Times*, June 26, 2012; Chris Beam, "The Definitive Guide to Butler Sex," *Bwog* (blog), reposted 2010, bwog.com/2010/09/01/the-definitive-guide-to-butler -sex-2/. For sex in airplane bathrooms, Brian Palmer, "'The Captain Requests That All Zippers Be Returned to the Upright Position,'" *Slate*, September 12, 2011, slate.com /articles/news_and_politics/explainer/2011/09/the_captain_requests_that_all_zippers _be_returned_to_the_upright_position.html; "The Germiest Places in America," *Health*, October 3, 2010, health.com/health/article/0,,20410740,00.html; and Kristine Fellizar, "What Happens If You Get Caught Having Sex on an Airplane?" *Bustle*, October 15, 2015, bustle.com/articles/114366-what-happens-if-you-get-caught-having-sex -on-an-airplane. Dr. Melvin Lansky's quote comes from Samuel G. Freedman, "Dreamed You Never Studied? Be Proud," *New York Times*, September 8, 2004.

The lemming footnote is from "Lemming Suicide Myth" by Riley Woodford, adfg.alaska.gov/index.cfm?adfg=wildlifenews.view_article&articles_id=560.

For my pages on how microbes help animals adapt, I used these sources, among others: *The Hidden Half of Nature* by David R. Montgomery and Anne Biklé; Frank Ryan, *Darwin's Blind Spot: Evolution beyond Natural Selection* (Boston: Houghton Mifflin Company, 2002); Elio Schaechter, *In the Company of Microbes*; V. G. Martinson et al., "A Simple and Distinctive Microbiota Associated with Honey Bees and Bumble Bees," *Molecular Ecology* 20, no. 3 (2011): 619–28; Patrick Monahan, "Bacteria Give Bird Its Sexy Smells," *Science*, June 26, 2016, sciencemag.org/news/2016/06/bacteria-give -bird-its-sexy-smells; J. Dylan Shropshire and Seth R. Bordenstein, "Speciation by Symbiosis: The Microbiome and Behavior," *mBio*, 7, no. 2 (2016): e01785-15, doi.org/10.1128/mBio.01785-15; Mindy Weisberger, "Human Gut Microbes Took Root before We Were Human," Live Science, July 21, 2016, livescience.com/55491-gut -microbes-predate-human-lineage.html; Claire Keeton, "Bacteria Give Each Meerkat Its Own Special Stink," *TimesLIVE*, June 12, 2017, timeslive.co.za/news/sci-tech/2017 -06-12-bacteria-give-each-meerkat-its-own-special-stink/; "Bacteria Produce Aphrodisiac That Sets Off Protozoan Mating Swarm," Phys.org, December 6, 2016, phys.org /news/2016-12-bacteria-aphrodisiac-protozoan-swarm.html; Alexandra A. Mushegian, "How Bacteria Induce Animal Metamorphosis," *Science Signaling* 9, no. 445 (2016): ec211, doi.org/10.1126/scisignal.aaj1824; Eugene Rosenberg and Ilana Zilber-Rosenberg, "Microbes Drive Evolution of Animals and Plants: The Hologenome Concept," *mBio* 7, no. 2 (2016): e01395-15, doi.org/10.1128/mBio.01395-15; Anna Azvolinsky, "Primates, Gut Microbes Evolved Together," *Scientist*, July 21, 2016, the-scientist.com/?articles .view/articleNo/46603/title/Primates—Gut-Microbes-Evolved-Together/; Andrew J. Tanentzap, Mark Vicari, and Dawn R. Bazely, "Ungulate Saliva Inhibits a Grass– Endophyte Mutualism," *Biology Letters* 10, no. 7 (2014): doi.org/10.1098/rsbl.2014.0460; Kevin D. Kohl et al., "Gut Microbes of Mammalian Herbivores Facilitate Intake of Plant Toxins," *Ecology Letters* 17, no. 10 (2014): 1238–46; and Ed Yong, "Gut Microbes Keep Species Apart," *Nature*, July 18, 2013, nature.com/news/gut-microbes-keep-species -apart-1.13408.

The footnote about changes in mouth bacteria was derived from "Diet Changes in Our Past Helped Harmful Microbes to Thrive," American Museum of Natural History Science Bulletins, April 2013, amnh.org/explore/science-bulletins/(watch)/human /news/diet-changes-in-our-past-helped-harmful-microbes-to-thrive. The notion that our microbial genes supersize us comes from Elizabeth A. Grice and Julia A. Segre, "The Human Microbiome: Our Second Genome," *Annual Review of Genomics and Human Genetics* 13 (2012): 151–70. The rancid smell of STDs is from Rachael Rettner, "Sexually Transmitted Disease Can Make Sweat Smell Foul," Live Science, December 9, 2011, livescience.com/17403-std-smell-gonorrhea.html. To that end, microbes are why we don't think our own farts stink, but we hate the smell of others': Rachel Feltman, "Why We Don't Think Our Own Farts Stink," *Washington Post*, November 10, 2014, washingtonpost.com/news/speaking-of-science/wp/2014/11/10/why-we-dont-think-our -own-farts-stink/?utm_term=.dadbd15f8035.

To read the entire story on the little mermaid kerfuffle at Columbia, originally aired January 11, 2002, go to thisamericanlife.org/radio-archives/episode/203/transcript. My paragraphs on Columbia's history of protest are built from the following: Kyla Bills, "Tracing the History of Student Activism & Why It's So Important Today," ://Milk.xyz, December 29, 2015, milk.xyz/feature/why-student-activism-remains-important/; Emily Jane Fox, "A Visual History of Campus Protests," *Vanity Fair*, November 20, 2015, vanityfair.com/news/photos/2015/11/campus-protest-photos; "Columbia 1968," columbia1968.com/history/; "Columbia University Students Win Divestment from Apartheid South Africa, United States, 1985," Global Nonviolent Action Database, nvdatabase.swarthmore.edu/content/columbia-university-students-win-divestment -apartheid-south-africa-united-states-1985; Aaron Holmes, "Columbia to Divest from Thermal Coal Producers," *Columbia Daily Spectator*, March 13, 2017, columbiaspectator .com/news/2017/03/13/columbia-to-divest-from-thermal-coal-producers/; Emily Bazelon, "Have We Learned Anything from the Columbia Rape Case?," *New York Times*, May 29, 2015; Joshua Barone, "Columbia Students Protest a Henry Moore Sculpture," *New York Times*, April 4, 2016; and Sasha Zients, "100 Camp Out on Low in Solidarity with CDCJ Occupation," *Columbia Daily Spectator*, April 22, 2016, columbiaspectator.com /news/2016/04/22/100-camp-out-low-solidarity-cdcj-occupation/.

Paragraphs about succession were augmented with information from Ed Smith and Jay Davison, "What Grows Back after the Fire?," Fact Sheet 96–40, University of

Nevada Cooperative Extension, anyflip.com/wine/kjho/basic; Noah Fierer et al., "Changes through Time: Integrating Microorganisms into the Study of Succession," *Research in Microbiology* 161, no. 8 (2010): 635–42; "Structure, Function and Diversity of the Healthy Human Microbiome," *Nature* 486 (2012): 207–14; and Elizabeth K. Costello et al., "The Application of Ecological Theory toward an Understanding of the Human Microbiome," *Science* 336, no. 6086 (2012): 1255–62. I read about the "greening of the self" in Joanna Macy, *Greening of the Self* (New York: Parallax Press, 2013), Kindle edition, and I found related ideas in Fritjof Capra, *The Web of Life: A New Scientific Understand of Living Systems* (New York: Anchor Books, 1996).

Chapter 10

I built this chapter from interviews with Kael Fischer of uBiota and the microbiologist Moselio Schaechter; an illuminating lecture by Maria Dominguez-Bello, "Factors Impacting the Infant Microbiome," at ASM Microbe 2016; the Rob Dunn Lab website: robdunnlab.com/projects/invisible-life/akkermansia/; and from reading the following books: Alanna Collen, *10% Human: How Your Body's Microbes Hold the Key to Health and Happiness* (New York: Harper, 2015); Martin J. Blaser, *Missing Microbes: How the Overuse of Antibiotics Is Fueling Our Modern Plagues* (New York: Henry Holt and Company, 2014), Kindle edition; Giulia Enders, *Gut: The Inside Story of Our Body's Most Underrated Organ* (London: Scribe, 2015); Marlene Zuk, *Riddled with Life: Friendly Worms, Ladybug Sex, and the Parasites That Make Us Who We Are* (Orlando: Harvest, 2007); Jessica Snyder Sachs, *Good Germs, Bad Germs: Health and Survival in a Bacterial World* (New York: Hill and Wang, 2007); Rob DeSalle and Susan L. Perkins, *Welcome to the Microbiome: Getting to Know the Trillions of Bacteria and Other Microbes In, On, and Around You* (New Haven: Yale University Press, 2015); Rob Dunn, *The Wild Life of Our Bodies: Predators, Parasites, and Partners That Shape Who We Are Today* (New York: HarperCollins, 2011); and *The Hidden Half of Nature* by David R. Montgomery and Anne Biklé, *I Contain Multitudes* by Ed Yong, and *The Amoeba in the Room* by Nicholas P. Money.

Dr. Constantine Sedikides was interviewed by Tim Adams, "Look Back in Joy: The Power of Nostalgia," *Guardian*, November 9, 2014, theguardian.com/society/2014/nov/09/look-back-in-joy-the-power-of-nostalgia. The Bored@Butler site is full of comments, and trawling through gives insight, of a sort, into Columbia culture. For some facts about the Dartmouth hangout, see Tyler Kingkade, "Why the Founder of the Notorious Bored@Baker Finally Shut This Site Down," *Huffington Post*, December 14, 2015, huffingtonpost.com/entry/bored-at-baker-shut-down_us_566efb7de4b0fcceee16f4270. News about Bored@Butler going offline can be found at bwog.com/2016/12/03/in-memoriam-of-boredbutler/; and the quote "judging a university . . . " can be found at talk.collegeconfidential.com/columbia-university/750450-bored-butler.html. I've never found "Schmasterpieces of Western Lit" again. Maybe someone decided, post-graduation, that it was better to take it offline.

Beside the books mentioned above, the paragraphs on the vaginome, pregnancy, and newborns were written using the following sources, among others: "Vaginal Bacteria Inhibit Sexual Transmission of HSV-2, Possibly Zika," Healio.com, June 3, 2017, healio.com/infectious-disease/emerging-diseases/news/online/%7B0c9e1d9d-87cb-4d52 -a931-fc444e22a450%7D/vaginal-bacteria-can-inhibit-sexual-transmission-of-zika-virus; Associated Press, "Women's Bacteria Thwarted Attempt at Anti-HIV Vaginal Gel," *STAT*, June 1, 2017, statnews.com/2017/06/01/womens-bacteria-hiv-vaginal-gel/; "Bacteria in the Cervix May Be Key to Understanding Premature Birth," Medical Xpress, January 27, 2017, medicalxpress.com/news/2017-01-bacteria-cervix -key-premature-birth.html; Chen Chen et al., "The Microbiota Continuum along Female Reproductive Tract and Its Relation to Uterine-Related Diseases," *Nature Communications* 8, no. 876 (2017): doi:10.1038/s41467-017-00901-0; Larry Forney, "The Intimate Relationship between Women & Their Bacteria," *Cultures*, accessed November 3, 2017, asmcultures.org/3-3/8/; Dani Cooper, "Exposure to Mother's Vaginal Fluids May Help Restore Bacteria of Babies Born via Caesarean," ABC Science, February 1, 2016, abc.net.au/news/science/2016-02-02/mums-vaginal-fluid-may-help-restore-bacteria-in -caesarean-babies/7130384; "Mother's Milk and the Infant Gut Microbiota: An Ancient Symbiosis," ScienceDaily, April 15, 2016, accessed August 13, 2017, sciencedaily.com/releases/2016/04/160415143610.htm; "Group B Strep Infection: GBS," American Pregnancy Association, accessed August 13, 2017, americanpregnancy.org /pregnancy-complications/group-b-strep-infection/; Holly Pevzner, "10 Things You Never Knew about Vernix," *Parents*, accessed August 13, 2017, parents.com/baby/care /skin/10-things-you-never-knew-about-vernix/; Leslie Burby, "101 Reasons to Breastfeed Your Child," ProMoM, accessed August 13, 2017, notmilk.com/101.html; Marie-Claire Arrieta et al., "Early Infancy Microbial and Metabolic Alterations Affect Risk of Childhood Asthma," *Science Translational Medicine* 7, no. 307 (2015): 307; Rebecca Dekker, "Evidence On: Erythromycin Eye Ointment for Newborns," Evidence Based Birth, August 3, 2017, evidencebasedbirth.com/is-erythromycin-eye-ointment-always -necessary-for-newborns/; Derrick M. Chu et al., "Maturation of the Infant Microbiome Community Structure and Function across Multiple Body Sites and in Relation to Mode of Delivery," *Nature Medicine* 23, no. 3 (2017): 314–26; and Michael Pollan, "Some of My Best Friends Are Germs," *New York Times*, May 15, 2013.

The oldest person in the world in 2016 ate two eggs a day. And cookies. Nick Allen, "World's Oldest Person, 116, Eats Diet of Bacon and Eggs," *Telegraph*, July 6, 2015, telegraph.co.uk/news/worldnews/northamerica/usa/11721581/Worlds-oldest-person-116 -eats-diet-of-bacon-and-eggs.html and Josh Clark, "How Can Some of the World's Oldest People Also Lead Unhealthy Lives?," HowStuffWorks, accessed August 13, 2017, health .howstuffworks.com/wellness/food-nutrition/diet-aging/oldest-people-unhealthy1.htm.

The paragraphs on the business of stool analysis were derived from the following: Ubiota.com; uBiome.com; Americangut.org; "My American Gut Individual Report," *Genomics, Etc.* (blog), 2013, cdwscience.blogspot.com/2013/11/my-american-gut -individual-report.html; Tina Hesman Saey, "Me and My Microbiome," *Science News*

185, no. 1 (2014): sciencenews.org/article/me-and-my-microbiome; Andrew Anthony, "I Had the Bacteria in My Gut Analysed. And This May Be the Future of Medicine," *Guardian*, February 11, 2014, theguardian.com/science/2014/feb/11/gut-biology-health -bacteria-future-medicine; Marcus J. Claesson et al., "Gut Microbiota Composition Correlates with Diet and Health in the Elderly," *Nature* 488 (2012): 178–84; Tanya Yatsunenko et al., "Human Gut Microbiome Viewed across Age and Geography," *Nature* 486 (2012): 222–27; "Here's the Poop on Getting Your Gut Microbiome Analyzed," *Gory Details* (blog), *Science News*, June 17, 2014, sciencenews.org/blog /gory-details/here%E2%80%99s-poop-getting-your-gut-microbiome-analyzed; Amy Talbott, "Microbiome Analysis: How One Tech Company Is Changing Health Behaviors, One Fecal Sample at a Time," TechRepublic, December 24, 2015, techrepublic.com /article/microbiome-analysis-how-one-tech-company-is-changing-health-behaviors -one-fecal-sample-at-a-time/; and "Personal Microbe Bank for Healthy Gut Bacteria," Springwise, November 24, 2015, accessed August 13, 2017, springwise.com/personal -microbe-bank-restore-healthy-gut-bacteria-antibiotics/.

Molecular anthropologist Christina Warinner presented her findings in her talk, "How 'Paleo' Is Your Diet?," at the American Museum of Natural History's SciCafe, New York City, April 28, 2016.

The very idea of cheating on a test was something I hadn't thought about in many years. To get up to speed, I used these resources: James M. Lang, *Cheating Lessons: Learning from Academic Dishonesty* (Cambridge: Harvard University Press, 2013), and Donald L. McCabe, Kenneth D. Butterfield, and Linda Klebe Treviño, *Cheating in College: Why Students Do It and What Educators Can Do about It* (Baltimore: Johns Hopkins University Press, 2012); Columbia's page "Dishonesty in Academic Work," college .columbia.edu/academics/academicdishonesty; the Educational Testing Service's page "Cheating Is a Personal Foul," glass-castle.com/clients/www-nocheating-org/adcouncil /research/cheatingfactsheet.html; Rebecca Martinson, "Take These Ingenious Ways Students Have Cheated on Tests and Definitely Don't Use Them on Your Finals *Wink,*" BroBible, accessed November 3, 2017, brobible.com/life/article/ingenious -ways-students-cheated/; Ariel Kaminer and Randy Leonard, "Reports of Cheating at Barnard College Cause a Stir," *New York Times*, May 8, 2013; James M. Lang, "How College Classes Encourage Cheating," *Boston Globe*, August 4, 2013, bostonglobe.com /ideas/2013/08/03/how-college-classes-encourage-cheating/3Q34x5ysYcplWNA3yO2eLK /story.html; Maryalene LaPonsie, "Top 10 Easiest and Most Unusual College Courses," *USA TODAY College*, October 12, 2011, college.usatoday.com/2011/10/12/top-10-easiest -and-most-unusual-college-courses/; "What Are Some Easy College Classes?," College Confidential, April 2010, talk.collegeconfidential.com/college-life/225895-what-are -some-easy-college-classes.html.

Information about all the proceeding microbiomes come from the books mentioned earlier and the following papers. For the mouth microbiome: Walter L. Loesche, "Chapter 99: Microbiology of Dental Decay and Periodontal Disease," in *Medical Microbiology*, 4th edition, ed. S. Baron (Galveston, TX: University of Texas Medical

Branch at Galveston, 1996) and J. Tonzetich, "Production and Origin of Oral Malodor: A Review of Mechanisms and Methods of Analysis," *Journal of Periodontology* 48, no. 1 (1977): 13–20. For the stomach microbiome: D. Y. Graham, "The Only Good *Helicobacter pylori* Is a Dead *Helicobacter pylori*," *Lancet* 350, no. 9070 (1997): 70–71; Nicholas Wade, "Ötzi the Iceman's Stomach Bacteria Offers Clues on Human Migration," *New York Times,* January 7, 2016; and DeAnna E. Beasley, "The Evolution of Stomach Acidity and Its Relevance to the Human Microbiome," *PLoS ONE* 10, no. 7 (2015): e0134116, doi.org/10.1371/journal.pone.0134116. For the paragraph on fecal transplants, I used the following sources: Belinda Smith, "Poo Transplants Transfer Much More Than Bacteria," *Cosmos,* July 13, 2016, cosmosmagazine.com/biology/poo-transplants-transfer-much-more-than-bacteria; Emily Eakin, "The Excrement Experiment," *New Yorker,* December 1, 2014; Joanna Lopez and Ari Grinspan, "Fecal Microbiota Transplantation for Inflammatory Bowel Disease," *Gastroenterology & Hepatology* 12, no. 6 (2016): 374–79; Carl Zimmer, "Fecal Transplants Can Be Life-Saving, but How?," *New York Times,* July 15, 2016; Diana P. Bojanova and Seth R. Bordenstein, "Fecal Transplants: What Is Being Transferred?," *PLoS Biology* 14, no. 7 (2016): e1002503, doi.org/10.1371/journal.pbio.1002503; and a talk by Ari Grinspan, a pioneer of fecal transplants, at Taste of Science in New York City, April 24, 2017. He described the German study that suggests phages may be the real heroes of fecal transplantation.

Regarding normal gut microbiota and human health, I used Seth R. Bordenstein and Kevin R. Theis, "Host Biology in Light of the Microbiome: Ten Principles of Holobionts and Hologenomes," *PLoS Biology* 13, no. 8 (2015): e1002226, doi.org/10.1371/journal.pbio.1002226; Jessica L. Mark Welch et al., "Spatial Organization of a Model 15-Member Human Gut Microbiota Established in Gnotobiotic Mice," *PNAS* 114, no. 43 (2017): E9105–14, doi.org/10.1073/pnas.1711596114; Catherine A. Lozupone et al., "Diversity, Stability and Resilience of the Human Gut Microbiota," *Nature,* 489 (2012): 220–30; Mahmoud Ghannoum, "The Mycobiome," *Scientist,* February 1, 2016, the-scientist.com/?articles.view/articleNo/45153/title/The-Mycobiome/; Sai Manasa Jandhyala et al., "Role of the Normal Gut Microbiota," *World Journal of Gastroenterology* 21, no. 29 (2015): 8787–803; and Rachel Lutz, "Antioxidant Metabolism Determined by Gut Bacteria," *MD Magazine,* January 14, 2016, mdmag.com/medical-news/antioxidant-metabolism-determined-by-gut-bacteria-.

Gut dysbiosis has been associated with many diseases, and new connections are announced every day, from "Scientists discover unsuspected bacterial link to bile duct cancer" to "Scientists detect gut bacteria in deepest reaches of failing lungs." I read many papers and articles of this sort, including Micah Manary, "How We Tell the Good Bacteria from the Bad," *Small Things Considered* (blog), 2011, schaechter.asmblog.org/schaechter/2011/07/how-we-tell-the-good-bacteria-from-the-bad.html?utm_source=feedburner&utm_medium=email&utm_campaign=Feed%3A+schaechter+%28Small+Things+Considered%29; Asa Hakansson and Goran Molin, "Gut Microbiota and Inflammation," *Nutrients* 3, no. 6 (2011): 637–82; N. A. Molodecky et al., "Increasing Incidence and Prevalence of the Inflammatory Bowel Diseases with Time, Based on

Systematic Review," *Gastroenterology* 142, no. 1 (2012): 46–54; "Early Studies Linking Gut Bacteria to Atherosclerosis Offer Tantalizing Glimpse at New Drug Target, Cleveland Clinic Reveals," BioSpace, December 22, 2015, biospace.com/News/early-studies-linking-gut-bacteria-to/403736/source=Featured; Ellen Kurek, "Role of Gut Bacteria in Asthma and Allergic Disease Becoming Clearer," *MD Magazine*, September 19, 2016, mdmag.com/medical-news/role-of-gut-bacteria-in-asthma-and-allergic-disease-becoming-clearer; and Diana Gitig, "Rise in Allergies Linked to War on Bacteria," *Wired*, March 28, 2012, wired.com/2012/03/allergic-bacteria-disease/.

The paragraphs on the gut–brain axis were augmented with material from the following articles and papers: Grace Niewijk, "A Psychobiotic Future?: The Mental Influence of Gut Bacteria," *The Scope* (blog), *Yale Scientific Magazine*, 2016, yalescientific.org/thescope/2016/12/a-psychobiotic-future-the-mental-influence-of-gut-bacteria/; "A Single Species of Gut Bacteria Can Reverse Autism-Related Social Behavior in Mice," Science Codex, June 16, 2016, sciencecodex.com/a_single_species_of_gut_bacteria_can_reverse_autismrelated_social_behavior_in_mice-184550; Peter Andrey Smith, "Can the Bacteria in Your Gut Explain Your Mood?" *New York Times*, June 23, 2015; R. H. Sandler et al., "Short-Term Benefit from Oral Vancomycin Treatment of Regressive-Onset Autism," *Journal of Child Neurology* 15, no. 7 (2000): 429–35; Marilia Carabotti et al., "The Gut-Brain Axis: Interactions between Enteric Microbiota, Central and Enteric Nervous Systems," *Annals of Gastroenterology* 28, no. 2 (2015): 203–209; and David Perlmutter, *Brain Maker: The Power of Gut Microbes to Heal and Protect Your Brain—For Life* (New York: Little, Brown and Company, 2015). I heard two talks at the ASM Microbe 2016 conference that were particularly insightful: Ira Blader, "*Toxoplasma gondii* Infections Alter Host Neuronal GABAergic Signaling" (which was about microbial interactions with brain and behavior) and Elaine Hsiao, "Microbiome-Nervous System Interactions in Health and Disease."

As further evidence that we should probably not assume nature is a bumbler and remove parts of ourselves, see Cassie Murdoch, "Oops! Looks Like You Might Need Your Appendix After All," *Jezebel*, March 5, 2012, jezebel.com/5890422/oops-looks-like-you-might-need-your-appendix-after-all, and Gene Y. Im et al., "The Appendix May Protect Against *Clostridium difficile* Recurrence," *Clinical Gastroenterology and Hepatology* 9, no. 12 (2011): 1072–77. Regarding the colonoscopy footnote, see Karen Weintraub, "Does Colon Cleansing Wipe Out Our Gut Microbiome?" *New York Times*, August 11, 2017.

Chapter 11

One of the many student chat groups in which I lurked was Giant Bomb, where I read the thread "Is there any greater anxiety than waiting for your grades?," giantbomb.com/forums/off-topic-31/is-there-any-greater-anxiety-than-waiting-for-your-497720/?page=2. Columbia students complain of tardy grades on *Bwog* (blog), bwog.com/2013/01/01/the-wall-of-shame-our-gpas-are-waiting/. Graduation types are well documented: Connie Chan, "The 11 Types of Graduating College Seniors," Her Campus, April 22, 2016, hercampus.com/life/campus-life/11-types-graduating-college-seniors. For the history of class rings, I read "The Weird and Cool History of Class Rings," Daniel's, accessed November 3, 2017, danielsjewelers.com/articles/The-weird-and-cool-history-of-Class-Rings.cfm, not the same place you get Columbia jewelry. That's cujewelry.com. For Columbia's history, I often visited wikicu.com. For my paragraphs on germophobes, I used Katherine Ashenburg, *The Dirt on Clean: An Unsanitized History* (New York: North Point Press, 2008); Sarah Zhang, "How 'Clean' Was Sold to America with Fake Science," Gizmodo, February 12, 2015, gizmodo.com/how-clean-was-sold-to-america-1685320177; Leanna Skarnulis, "Cleanliness Rules Germophobes' Lives," MedicineNet.com, accessed August 13, 2017, medicinenet.com/script/main/art.asp?articlekey=46748&page=1; and Mary Roach, "Germs, Germs Everywhere. Are You Worried? Get Over It," *New York Times*, November 9, 2004. Jonathan Eisen's quote comes from an essay he wrote, "Microbiomania and 'Overselling the Microbiome,'" *The Tree of Life* (blog), 2014, phylogenomics.blogspot.com/p/blog-page.html.

Expanding my list of surprising places where you may find bacteria (not so surprising once you know bacteria are everywhere) has been one of life's little pleasures. My favorite new places: birthday cakes (after blowing out the candles), Sarah Young, "Blowing Out Birthday Candles Increases Bacteria on Cake by 1,400% Study Reveals," *Independent*, July 31, 2017, independent.co.uk/life-style/blowing-out-birthday-candles-increases-cake-bacteria-1400-per-cent-saliva-spread-icing-study-reveals-a7868671.html; and your Sunday roast, Tom Rawstorne, "Dying for Your Sunday Roast: How the Lethal Bacteria That Can Lurk Inside Leads to 500,000 Cases of Food Poisoning a Year . . . And 100 Deaths," DailyMail.com, October 23, 2016, dailymail.co.uk/news/article-3865400/Dying-Sunday-roast-lethal-bacteria-lurk-inside-leads-500-000-cases-food-poisoning-year-100-deaths.html. The 5-second rule is bogus: Christopher Mele, "'Five-Second Rule' for Food on Floor Is Untrue, Study Says," *New York Times*, September 19, 2016, and Aaron E. Carroll, "I'm a Doctor. If I Drop Food on the Kitchen Floor, I Still Eat It.," *New York Times*, October 10, 2016.

Clair Folsome's quote comes from a thought experiment he posed in 1985. I read about it in *One Plus One Equals One* by John Archibald.

The skin microbiome is well reported in *Welcome to the Microbiome* by Rob DeSalle and Susan L. Perkins. Additional facts used in these paragraphs were drawn from: A. L. Cogen, V. Nizet, and R. L. Gallo, "Skin Microbiota: A Source of Disease or Defence?," *British Journal of Dermatology* 158, no. 3 (2008): 442–55; Noah Fierer et al., "The Influence of Sex, Handedness, and Washing on the Diversity of Hand Surface Bacteria," *PNAS* 105, no. 46 (2008): 17994–99; "The Microbiome of Your Skin," American Museum of Natural History, accessed August 13, 2017, amnh.org/exhibitions/the-secret-world-inside-you/the-microbiome-of-your-skin; Marcus H. Y. Leung, Kelvin C. K. Chan, and Patrick K. H. Lee, "Skin Fungal Community and Its Correlation with Bacterial Community of Urban Chinese Individuals," *Microbiome* 4 (2016): 46, doi.org/10.1186/s40168-016-0192-z; Melanie Hemsworth, "Skin Bacteria Safeguard

Babies from Developing Eczema, Irish Scientists Learned," Parent Herald, October 9, 2017, parentherald.com/articles/72112/20161009/health-news-update-skin-bacteria -safeguard-babies-from-developing-eczema-researches-learned.htm; Alexandra Ossola, "Can Bacteria Make Your Skin Healthier?," Daily Beast, August 18, 2016, thedailybeast.com/can-bacteria-make-your-skin-healthier; Sara Ivry, "That Fresh Feeling," *New York Times*, December 16, 2007; Kate Lunau, "I Skipped Showering for Two Weeks and Bathed in Bacteria Instead," *Motherboard*, June 17, 2016, motherboard.vice.com /en_us/article/78k4n4/showering-and-bathed-in-bacteria-instead-mother-dirt-aobiome -microbiome; "'Acne Bacteria' Integral Part of Skin Microbiota Defense System," *Genetic Engineering and Biotechnology News*, November 14, 2016, genengnews.com /gen-news-highlights/acne-bacteria-integral-part-of-skin-microbiota-defense-system /81253430; Andy Coghlan, "How Lack of Oxygen Makes Bacteria Cause Acne and How to Stop It," *New Scientist*, October 28, 2016, newscientist.com/article/2110826-how -lack-of-oxygen-makes-bacteria-cause-acne-and-how-to-stop-it/; Adam Friedman, "Acne Treatment: Antibiotics Don't Need to Kill Bacteria to Clear Up Your Skin," The Conversation, March 18, 2016, theconversation.com/acne-treatment-antibiotics-dont -need-to-kill-bacteria-to-clear-up-your-skin-56188; Arielle R. Nagler, Emily C. Milam, and Seth J. Orlow, "The Use of Oral Antibiotics Before Isotretinoin Therapy in Patients with Acne," *Journal of the American Academy of Dermatology* 74, no. 2 (2016): 273–79; Anna Azvolinsky, "Birth of the Skin Microbiome," *Scientist*, November 17, 2015, the-scientist.com/?articles.view/articleNo/44488/title/Birth-of-the-Skin-Microbiome/; Anthony M. Cundell, "Microbial Ecology of the Human Skin," *Microbial Ecology* (2016): doi.org/10.1007/s0024; Louise Gagnon, "Knowledge of Skin Microbiome Provides Insights into Dermatological Conditions," *Dermatology Times*, November 30, 2016, dermatologytimes.modernmedicine.com/dermatology-times/news/%20title -raw%5D-7; Poncie Rutsch, "Meet the Bacteria That Make a Stink in Your Pits," NPR, March 31, 2015, npr.org/sections/health-shots/2015/03/31/396573607/meet-the-bacteria -that-make-a-stink-in-your-pits; Jessica Hamzelou, "Stinky Armpits? Bacteria from a Less Smelly Person Can Fix Them," *New Scientist*, February 10, 2017, newscientist.com /article/2120923-stinky-armpits-bacteria-from-a-less-smelly-person-can-fix-them/; Associated Press, "Scientists Mix Friendly Bacteria into Lotions to Ward Off Bad Germs," CBS News, February 22, 2017, cbsnews.com/news/scientists-mix-friendly -bacteria-into-lotions-to-ward-off-bad-germs/; Julie Horvath, "Antiperspirant Alters the Microbial Ecosystem on Your Skin," news.ncsu.edu, February 2, 2016, news.ncsu .edu/2016/02/dunn-armpit-2016/; Lydia Ramsey and Kevin Loria, "Some People Are More Likely to Get Bitten by Mosquitoes—Here's Why," Business Insider, February 24, 2016, businessinsider.com/skin-bacteria-can-deter-mosquito-bites-2016-2; Evan Anstey, "Research Shows Not Shaving Can Help Prevent Bacteria, Infections," WIVB.com, January 22, 2016, wivb.com/2016/01/22/research-shows-not-shaving-can-help -prevent-bacteria-infections/; Julia Scott, "My No-Soap, No-Shampoo, Bacteria-Rich Hygiene Experiment," *New York Times*, May 22, 2014; "Mother Dirt: Bringing Back the Good Bacteria," Vimeo video, 1:57, posted by Rodman Media, vimeo.com/180930324;

Elise Minton, "Bacteria Could Be the Answer to Perfect Skin," *New Beauty,* December 27, 2015, newbeauty.com/blog/dailybeauty/9104-is-bacteria-the-answer-to-perfect-skin/; Danielle Fontana, "Bacteria-Infused Skin Care You Need to Get Your Hands On," *New Beauty,* April 12, 2016, newbeauty.com/slideshow/2205-probiotic-infused-skincare/12/; "Immune System Helps Dictate Microbes Found on Human Skin," *Genetic Engineering & Biotechnology News,* October 30, 2013, genengnews.com/gen-news-highlights /immune-system-helps-dictate-microbes-found-on-human-skin/81249042; Joshua A. Krisch, "Next Generation: Personalized Probiotic Skin Care," *Scientist,* February 27, 2017, the-scientist.com/?articles.view/articleNo/48628/title/Next-Generation— Personalized-Probiotic-Skin-Care/; and Ed Yong, "A Probiotic Skin Cream Made with a Person's Own Microbes," *Atlantic,* February 22, 2017, theatlantic.com/science /archive/2017/02/a-personalized-probiotic-skin-cream-made-with-a-persons-own -microbes/517473/.

Probiotics are ripe with potential and rife with bull. I started off reading Case Adams, *Probiotics: Protection against Infection: Using Nature's Tiny Warriors to Stem Infection and Fight Disease* (Wilmington, DE: Logical Books, 2012). And I went on to use the following: "Could Probiotics Be the Next Big Thing in Acne and Rosacea Treatments," American Academy of Dermatology, January 30, 2014, aad.org/media/news -releases/could-probiotics-be-the-next-big-thing-in-acne-and-rosacea-treatments; April Long, "Why You Should Put Yogurt on Your Face," *Elle,* March 10, 2016, elle.com /beauty/makeup-skin-care/a33771/probiotic-skincare/; Moises Velasquez-Manoff, "Microbes, a Love Story," *New York Times,* February 10, 2017; Cari Nierenberg, "4 Conditions Probiotics Are Likely to Treat," Live Science, June 11, 2014, livescience .com/46768-probiotics-gut-conditions-treat.html; Lydia Ramsey, "You've Been Fighting Morning Breath All Wrong," *Business Insider,* November 14, 2015, businessinsider.com /can-you-use-a-probiotic-to-fix-bad-breath-2015-11; M. C. Mekkes et al., "The Development of Probiotic Treatment in Obesity: A Review," *Beneficial Microbes* 5, no. 1 (2014): 19–28; Theodoros Kelesidis and Charalabos Pothoulakis, "Efficacy and Safety of the Probiotic *Saccharomyces boulardii* for the Prevention and Therapy of Gastrointestinal Disorders," *Therapeutic Advances in Gastroenterology* 5, no. 2 (2012): 111–25; S. E. Gilliland, "Health and Nutritional Benefits from Lactic Acid Bacteria," *FEMS Microbiology Reviews* 7, no. 1–2 (1990): 175–88; "First Probiotic to Address the Critical Role of Fungi in Digestive Health Hits Market," PR Newswire, March 7, 2017, prnewswire.com /news-releases/first-probiotic-to-address-the-critical-role-of-fungi-in-digestive-health -hits-market-300419148.html; A. Homayouni et al., "Effects of Probiotics on the Recurrence of Bacterial Vaginosis: A Review," *Journal of Lower Genital Tract Disease* 18, no. 1 (2014): 79–86; Nicholas Bakalar, "Probiotics Are Common in Hospitals, but Evidence Is Lacking," *Well* (blog), *New York Times,* 2016, well.blogs.nytimes.com/2016/03/03 /probiotics-are-common-in-hospitals-but-evidence-is-lacking/; "Which 2017 Probiotic Supplements Are the Most Effective?," Smarter Reviews, accessed August 14, 2017, smarter-reviews.com/lp/sr-probiotics?tr=lg155&gclid=Cj0KCQjwlMXMBRC1ARIsAK KGuwhQ50wut5RxAZB0vhBqpjMpf9_3WNtcgj12BktstLxBiWTL5vD_AkcaAierEALw_ wcB; *Guidelines for the Evaluation of Probiotics in Food*, Joint FAO/WHO Working Group Report, 2002, who.int/foodsafety/fs_management/en/probiotic_guidelines .pdf?ua=1; "Probiotics Market Analysis by Application (Probiotic Foods & Beverages (Dairy Products, Non-Dairy Products, Cereals, Baked Food, Fermented Meat Products, Dry Food), Probiotic Dietary Supplements (Food Supplements, Nutritional Supplements, Specialty Nutrients, Infant Formula), Animal Feed Probiotics), by End-Use (Human Probiotics, Animal Probiotics) and Segment Forecast to 2024," Grand View Research, September 2016, accessed August 14, 2017, grandviewresearch.com/industry -analysis/probiotics-market; Andrea Cespedes, "How Much Yogurt Do You Need for Probiotics?," Livestrong.com, October 3, 2017, livestrong.com/article/530753-how-much-yogurt-do-you-need-for-probiotics/; Tony Whitfield, "Healthy People Who Drink Probiotic Drinks Full of 'Good Bacteria' Wasting Their Time, According to Report," *Mirror*, May 10, 2016, mirror.co.uk/news/world-news/healthy-people-who-drink-probiotic-7935413; Beth Skwarecki, "Yoghurt Isn't Always the Best Source of Probiotic Bacteria," Lifehacker Australia, December 13, 2015, lifehacker.com.au/2015/12/ yoghurt-isnt-always-the-best-source-of-probiotic-bacteria/; Ryan F. Mandelbaum, "Are Vitamin Supplements Killing Our Gut Bacteria?," Gizmodo, May 17, 2017, gizmodo .com/are-vitamin-supplements-killing-our-gut-bacteria-1795299961; and "Discover the Digestive Benefits of Fermented Foods," *Tufts University Health & Nutrition Letter*, February 2014, nutritionletter.tufts.edu/issues/10_2/current-articles/Discover-the -Digestive-Benefits-of-Fermented-Foods_1383-1.html.

On prebiotics, fiber, and diets, I found the following books particularly helpful: Raphael Kellman, *The Microbiome Diet: The Scientifically Proven Way to Restore Your Gut Health and Achieve Permanent Weight Loss* (Boston: Da Capo Press, 2014); Jeff Leach, *Rewild: You're 99% Microbe. It's Time You Started Eating Like It* (CreateSpace Independent Publishing Platform, 2015); *Missing Microbes* by Martin J. Blaser; *Welcome to the Microbiome* by Rob DeSalle and Susan L. Perkins; and *The Hidden Half of Nature* by David R. Montgomery and Anne Biklé. The papers and articles I used to write these paragraphs are Joanna Slavin, "Fiber and Prebiotics: Mechanisms and Health Benefits," *Nutrients* 5, no. 4 (2013): 1417–35; Hannah D. Holscher, "Dietary Fiber and Prebiotics and the Gastrointestinal Microbiota," *Gut Microbes* 8, no. 2 (2017): 172–84; Amy M. Brownawell et al., "Prebiotics and the Health Benefits of Fiber: Current Regulatory Status, Future Research, and Goals," *Journal of Nutrition* 142, no. 5 (2012): 962–74; Peter J. Turnbaugh et al., "The Effect of Diet on the Human Gut Microbiome: A Metagenomic Analysis in Humanized Gnotobiotic Mice," *Science Translational Medicine* 1, no. 6 (2009): doi.org/10.1126/scitranslmed.3000322; Jef Akst, "Gut Bacteria Vary with Diet," *Scientist*, December 13, 2013, the-scientist.com/?articles.view/articleNo/38643/title /Gut-Bacteria-Vary-with-Diet/; Claudia Wallis, "How Gut Bacteria Help Make Us Fat and Thin," *Scientific American*, June 1, 2014, scientificamerican.com/article/how-gut -bacteria-help-make-us-fat-and-thin/; "Which Foods Can Improve Your Gut Bacteria?" BBC, January 30, 2017, bbc.com/news/health-38800977; Mary Jane Brown, "How Short-Chain Fatty Acids Affect Health and Weight," Healthline, April 2, 2016, healthline.com

/nutrition/short-chain-fatty-acids-101; Roni Caryn Rabin, "A Gut Makeover for the New Year," *Well* (blog), *New York Times*, 2016, nytimes.com/2016/12/29/well /eat/a-gut-makeover-for-the-new-year.html?em_pos=medium&emc=edit_sc_20170102&nl =science-times&nl_art=1&nlid=54784482&ref=headline&te=1&_r=0; Peter J. Turnbaugh, "A Core Gut Microbiome in Obese and Lean Twins," *Nature* 457 (2009): 480–84; and Michaeleen Doucleff, "Chowing Down on Meat, Dairy Alters Gut Bacteria a Lot, and Quickly," *The Salt*, 2013, npr.org/sections/thesalt/2013/12/10/250007042/chowing -down-on-meat-and-dairy-alters-gut-bacteria-a-lot-and-quickly.

The tendency of some bacteria to collect at disease sites is well reported. I referred to Anne Trafton, "Cancer-Fighting Bacteria: Engineers Program *E. coli* to Destroy Tumor Cells," MIT News, July 20, 2016, news.mit.edu/2016/cancer-fighting-bacteria -0720; Knvul Sheikh, "The Breast Has Its Own Microbiome—and the Mix of Bacteria Could Prevent or Encourage Cancer," *Scientific American*, October 1, 2016, scientificamerican.com/article/the-breast-has-its-own-microbiome-and-the-mix-of -bacteria-could-prevent-or-encourage-cancer/; and American Society for Microbiology, "Beneficial Bacteria May Protect Breasts from Cancer," EurekAlert!, June 24, 2016, eurekalert.org/pub_releases/2016-06/asfm-bbm062216.php. Mark D. Vincent's super interesting quote on the microbial character of cancer cells is from Marcia Stone, "Oddly Microbial: Cancer Cells," *Small Things Considered* (blog), 2012, schaechter .asmblog.org/schaechter/2012/06/oddly-microbial-cancer-cells.html.

Regarding the recruitment of bacteria for cancer therapy, I used these sources: "New Salmonella Strain Helps Fight Cancer," YouTube video, 2:06, posted by CBS New York, youtube.com/watch?v=7XKQ-AxzVBE; "'Bugs' as Drugs: Harnessing Novel Gut Bacteria for Human Health," ScienceDaily, May 4, 2016, accessed August 14, 2017, sciencedaily.com/releases/2016/05/160504141127.htm; Rob Matheson, "Reprogramming Gut Bacteria as 'Living Therapeutics,'" *Bioscience Technology*, April 6, 2016, biosciencetechnology.com/news/2016/04/reprogramming-gut-bacteria-living-therapeutics; Kate Yandell, "Microbes Meet Cancer," *Scientist*, April 1, 2016, the-scientist .com/?articles.view/articleNo/45616/title/Microbes-Meet-Cancer/; Sally Adee, "Self-Destructing Bacteria Are Engineered to Kill Cancer Cells," *New Scientist*, July 20, 2016, newscientist.com/article/2098172-self-destructing-bacteria-are-engineered-to-kill-cancer-cells/; "Researchers Engineer Bacteria to Hunt Down Pathogens in the Gut," Fischell Department of Bioengineering, University of Maryland, May 26, 2017, bioe .umd.edu/news/news_story.php?id=10599; Gina Kolata, "Could Alzheimer's Stem from Infections? It Makes Sense, Experts Say," *New York Times*, May 25, 2016; Rachael Rettner, "People with Alzheimer's May Have More Bacteria in Their Brains," Live Science, July 18, 2017, livescience.com/59850-alzheimers-disease-bacteria-brain.html; and David C. Emery et al., "16S rRNA Next Generation Sequencing Analysis Shows Bacteria in Alzheimer's Post-Mortem Brain," *Frontiers in Aging Neuroscience* (2017): doi.org /10.3389/fnagi.2017.00195.

Flesh-eating bacteria disease is rare: "Necrotizing Fasciitis," Centers for Disease Control and Prevention, updated July 3, 2017, accessed November 5, 2017, cdc.gov

/features/necrotizingfasciitis/. On the other hand, H. P. Weil, U. Fischer-Brügge, and P. Koch, "Potential Hazard of Wound Licking," *New England Journal of Medicine* 346 (2002): 1336. The toxoplasmosis story is a bizarre one, well-described by Kathleen McAuliffe, "How Your Cat Is Making You Crazy," *Atlantic*, March 2012, theatlantic .com/magazine/archive/2012/03/how-your-cat-is-making-you-crazy/308873/. Other sources for this paragraph are Carl Zimmer, "Parasites Practicing Mind Control," *New York Times*, August 28, 2014; Ed Yong, "How the Zombie Fungus Takes Over Ants' Bodies to Control Their Minds," *Atlantic,* November 14, 2017, theatlantic.com/science /archive/2017/11/how-the-zombie-fungus-takes-over-ants-bodies-to-control-their-minds /545864/; and my own *Mycophilia.*

On the subject of antibiotic-resistant pathogens, *Missing Microbes* by Martin J. Blaser provided an overview. More specifically, I used the following sources, among others: Melinda Wenner Moyer, "The Looming Threat of Factory-Farm Superbugs," *Scientific American*, December 2016; Donald G. McNeil Jr., "Deadly, Drug-Resistant 'Superbugs' Pose Huge Threat, W.H.O. Says," *New York Times*, February 27, 2017; "Science of Resistance: Antibiotics in Agriculture," Alliance for the Prudent Use of Antibiotics, accessed August 14, 2017, emerald.tufts.edu/med/apua/about_issue /antibiotic_agri.shtml; Sabrina Tavernise, "Antibiotic-Resistant Infections Lead to 23,000 Deaths a Year, C.D.C Finds," *New York Times*, September 16, 2013; Sabrina Tavernise, "U.S. Aims to Curb Peril of Antibiotic Resistance," *New York Times*, September 18, 2014; Aaron E. Carroll, "We're Losing the Race Against Antibiotic Resistance, but There's Also Reason for Hope," *New York Times*, March 7, 2016; "The Spread of Superbugs," *Economist*, March 31, 2011, economist.com/node/18483671; Review on Antimicrobial Resistance, *Securing New Drugs for Future Generations: The Pipeline of Antibiotics*, May 2015, https://amr-review.org/sites/default/files/SECURING%20NEW%20 DRUGS%20FOR%20FUTURE%20GENERATIONS%20FINAL%20WEB_0.pdf; Raffi Khatchadourian, "The Unseen: Millions of Microbes Are Yet to Be Discovered. Will One Hold the Ultimate Cure?," *New Yorker*, June 20, 2016; Ezekiel J. Emanuel, "How to Develop New Antibodies," *New York Times*, February 24, 2015; Jessica Calefati, "Antibiotics Ban: California First State to Outlaw Routine Use of Bacteria-Fighting Drugs in Livestock," Mercury News, October 10, 2015, mercurynews.com/2015/10/10/antibiotics -ban-california-first-state-to-outlaw-routine-use-of-bacteria-fighting-drugs-in-livestock/; Amy Mayer, "Feeding Bacteria to Livestock Could Cut Antibiotic Use," Iowa Public Radio, April 11, 2016, iowapublicradio.org/post/feeding-bacteria-livestock -could-cut-antibiotic-use#stream/0; Kalyan Kumar, "Air Pollution Discovered as a Means of Transmission for Antibiotic-Resistant Bacteria," *Tech Times*, November 21, 2016, techtimes.com/articles/186552/20161121/air-pollution-discovered-as-a-means-of -transmission-for-antibiotic-resistant-bacteria.htm; Eric Boodman, "Bacteria or Virus? New Diagnostic Tool Could Curb Antibiotic Overuse," *STAT*, January 20, 2016, statnews.com/2016/01/20/diagnostic-bacteria-virus-antibiotics/; Marissa Fessenden, "Future Antibiotics for Humans Could Come from Ant Fungus Gardens," Smithsonian .com, August 17, 2015, smithsonianmag.com/smart-news/new-antibiotics-humans

-could-come-ant-fungus-gardens-180956300/?no-ist; Cameron Phelps, "Tasmanian Devil's Milk Can KILL the Most Deadly Drug-Resistant Bacteria Known—Including Golden Staph," DailyMail.com, October 17, 2016, dailymail.co.uk/news/article -3842476/Tasmanian-devil-milk-kill-deadly-drug-resistant-bacteria-known-including -golden-staph.html; Beth Ferry et al., "Don't Wait, Be Ready! New Antibiotic Rules for 2017," Michigan State University Extension, msue.anr.msu.edu/news/dont_wait_be _ready_new_antibiotic_rules_for_2017; and Sabrina Tavernise, "F.D.A. Bans Sale of Many Antibacterial Soaps, Saying Risks Outweigh Benefits," *New York Times*, September 2, 2016.

Here's the report on the bit of mold that sold for $14,617: Associated Press, "Old Mold from Penicillin Discoverer Auctioned for $14,617," *New York Times*, March 6, 2017.

You can watch Representative Waters of California be threatened with a presentation of the mace at "Controversy on House Floor—Request to Present Mace," C-SPAN video, 1:35, posted by RXB, www.c-span.org/video/?c3342940/controversy-house-floor -request-present-mace. The hoopla about Obama speaking at Barnard's commencement can be read on *Bwog* (blog), bwog.com/2012/03/03/breaking-obama-to-speak-at -barnards-commencement/#comment-347741.

Chapter 12

The ASM Microbe 2016 conference paragraphs include information from Alan Boyle, "Gates Foundation Commits $100M to White House's National Microbiome Initiative," GeekWire, May 13, 2016, geekwire.com/2016/gates-foundation-commits -100m-white-houses-national-microbiome-initiative/; Ed Yong, "The White House Launches the National Microbiome Initiative," *Atlantic*, May 13, 2016, theatlantic.com /science/archive/2016/05/white-house-launches-the-national-microbiome-initiative /482598/; and Christopher Durso, "Lessons Learned from Combining Two Meetings," *Convene*, June 20, 2016, pcmaconvene.org/places-spaces/destinations/lessons-learned -from-combining-two-meetings/.

The papers and articles I used to write the pages on microbial clouds and indoor microbial populations were sourced from Noah Fierer et al., "Forensic Identification Using Skin Bacterial Communities," *PNAS* 107, no. 14 (2010): 6477–81; James F. Meadow et al., "Humans Differ in Their Personal Microbial Cloud," *PeerJ* 3 (2015): e1258, doi .org/10.7717/peerj.1258; W. C. Noble et al., "Quantitative Studies on the Dispersal of Skin Bacteria into the Air," *Journal of Medical Microbiology* 9, no. 1 (1976): 53–61; Denina Hospodsky et al., "Human Occupancy as a Source of Indoor Airborne Bacteria," *PLoS ONE* 7, no. 4 (2012): e34867, doi.org/10.1371/journal.pone.0034867. James F. Meadow et al., "Significant Changes in the Skin Microbiome Mediated by the Sport of Roller Derby," *PeerJ* 1 (2013): e53, doi.org/10.7717/peerj.53; Simon Lax et al., "Longitudinal Analysis of Microbial Interaction between Humans and the Indoor Environment," *Science* 345, no. 6200 (2014): 1048–52; "The Home Microbiome Project," YouTube video, 2:32, posted by Argonne National Laboratory, youtube.com /watch?v=dQCBpmUZlF4; Steven W. Kembel et al., "Architectural Design Influences the Diversity and Structure of the Built Environment Microbiome," *ISME Journal* 6 (2012): 1469–79; Rachel I. Adams et al., "Chamber Bioaerosol Study: Outdoor Air and Human Occupants as Sources of Indoor Airborne Microbes," *PLoS ONE* 10, no. 5 (2015): e0128022, doi.org/10.1371/journal.pone.0128022; Albert Barberán et al., "The Ecology of Microscopic Life in Household Dust," *Proceedings of the Royal Society B* 282, no. 1814 (2015): doi.org/10.1098/rspb.2015.1139; Robert R. Dunn et al., "Home Life: Factors Structuring the Bacterial Diversity Found within and between Homes," *PLoS ONE* 8, no. 5 (2013): e64133, doi.org/10.1371/journal.pone.0064133; Gilberto E. Flores et al., "Diversity, Distribution and Sources of Bacteria in Residential Kitchens," *Environmental Microbiology* 15, no. 2 (2013): 588–96; "Common Antibacterial Triclosan Found in Most Freshwater Streams," ScienceDaily, May 25, 2016, sciencedaily.com/releases /2016/05/160525121602.htm; Coco Ballantyne, "Strange but True: Antibacterial Products May Do More Harm Than Good," *Scientific American*, June 7, 2007, scientificamerican .com/article/strange-but-true-antibacterial-products-may-do-more-harm-than-good/; Kei E. Fujimura et al., "Man's Best Friend? The Effect of Pet Ownership on House Dust Microbial Communities," *Journal of Allergy and Clinical Immunology* 126, no. 2 (2010): 410–12; "Allergy Facts and Figures," Asthma and Allergy Foundation of America, accessed November 5, 2017, aafa.org/page/allergy-facts.aspx; "Facts about *Stachybotrys chartarum* and Other Molds," Centers for Disease Control and Prevention, accessed November 5, 2017, cdc.gov/mold/stachy.htm; Ronald E. Gots, "Correcting Mold Misinformation," Harvard University, Department of Physics, users.physics.harvard .edu/~wilson/soundscience/mold/gots1.html; "The Key to Mold Control Is Moisture Control," US Environmental Protection Agency, accessed November 5, 2017, epa.gov /mold; Jordan Smith, "The 'Mold Queen' Fights Back," *Austin Chronicle*, March 21, 2003, austinchronicle.com/news/2003-03-21/150675/; John Chase et al., "Geography and Location Are the Primary Drivers of Office Microbiome Composition," *mSystems* 1, no. 2 (2016): e00022-16, doi.org/10.1128/mSystems.00022-16; and Albert Barberán et al., "Continental-Scale Distributions of Dust-Associated Bacteria and Fungi," *PNAS* 112, no. 18 (2015): 5756–61. Donald MacRae's quote comes from an essay he wrote, "Darwinism and the Social Sciences," in S.A. Barnett, ed., *A Century of Darwin* (Cambridge, MA: Harvard University Press, 1958), but I read about it in Robert Ardrey, *The Territorial Imperative: A Personal Inquiry into the Animal Origins of Property and Nations* (New York: Dell, 1966).

See the Bill Gates keynote speech at ASM Microbe 2016: "Bill Gates Keynote at ASM MICROBE 2016," YouTube video, 57:40, posted by Expovista TV, youtube.com /watch?v=D2pWH6hDZHM; for the footnote on *S. marcescens*, F. Aragona, "Bartolomeo Bizio and the Phenomenon of 'Purpurine Polenta': Unknown History of *Serratia marcescens* Infections," *Archivos Españoles Urología* 5, no. 2 (1999): 105–11; Jennifer Frazer, "Miraculous Microbes: They Make Holy Statues 'Bleed'—and Can Be Deadly, Too," *Scientific American*, November 11, 2011, scientificamerican.com/article/serratia -marcescens-bacteria-holy-statues-bleed/; and Martin A. Lee and Bruce Shlain, *Acid Dreams: The Complete Social History of LSD: The CIA, the Sixties, and Beyond* (New

York: Grove Press, 1985). To see the singing microbiologist, go to the *MicrobeWorld Video* podcast, "Sheldon Campbell Is: The Singing Microbiologist," 6:20, American Society for Microbiology, asm.org/index.php/podcasts/microbeworld-video/item /2951-mwv-episode-81-sheldon-campbell-the-singing-microbiologist. The books I used for this section were *Welcome to the Microbiome* by Rob DeSalle and Susan L. Perkins, *Deadly Companions* by Dorothy H. Crawford, and *Mycophilia*.

The pages on microbial meteorology were based on Dr. Cindy Morris's talk, "Friends or Foes? When Plant Pathogens Make Rain," about *P. syringae* at the ASM Microbe 2016 conference; *March of the Microbes* by John L. Ingraham; and *The Amoeba in the Room* by Nicholas P. Money. I also used the following articles and papers: Chelsea Wald, "The Surprising Importance of Stratospheric Life," *Nautilus*, June 2, 2016, nautil .us/issue/37/currents/the-surprising-importance-of-stratospheric-life; U. Pöschl et al., "Rainforest Aerosols as Biogenic Nuclei of Clouds and Precipitation in the Amazon," *Science* 329, no. 5998 (2010): 1513–16; Luis Pérez Vicente, "Evolution of *Fusarium* Taxonomy," Food and Agriculture Organization of the United Nations, accessed November 5, 2017, fao.org/fileadmin/templates/agphome/documents/Pests_Pesticides /caribbeantr4/06Evolution.pdf; Cindy E. Morris and David C. Sands, *From Grains to Rain: The Link between Landscape, Airborne Microorganisms and Climate Processes*, e-book, accessed August 14, 2017, bioice.files.wordpress.com/2012/05/grainsrain _v26apr2012d.pdf; "The Rainmaker Named Sue: David Sands at TEDxBozeman," You-Tube video, 10:22, posted by TEDx Talks, youtube.com/watch?v=_9ZeYxoWsuk; Ravindra Pandey et al., "Ice-Nucleating Bacteria Control the Order and Dynamics of Interfacial Water," *Science Advances* 2, no. 4 (2016): e1501630, doi.org/10.1126 /sciadv.1501630; "Snomax Sno Inducer," YouTube video, 1:31, posted by Snomax LLC, youtube.com/watch?v=SWa17IQdE-8; Brent C. Christner, "Cloudy with a Chance of Microbes," *Microbe* 7, no. 2 (2012): 70–75; Robert M. Bowers et al., "Characterization of Airborne Microbial Communities at a High-Elevation Site and Their Potential to Act as Atmospheric Ice Nuclei," *Applied and Environmental Microbiology* 75, no. 15 (2009): 5121–30; National Research Council, *Critical Issues in Weather Modification Research* (Washington, DC: National Academies Press, 2003); Stephanie Warren, "Bacteria Live at 33,000 Feet," *Popular Science*, July 2013; Daniella Gat, "Origin-Dependent Variations in the Atmosphere Microbiome Community in Eastern Mediterranean Dust Storms," *Environmental Science & Technology* 51, no. 12 (2017): 6709–18; Elio Schaechter, "Raindrops Keep Falling on Their Heads (Thanks to Mushroom Spores, That Is)," *Small Things Considered* (blog), 2016, schaechter.asmblog.org/schaechter/2016/01/raindrops -keep-falling-on-their-heads-thanks-to-mushroom-spores-that-is.html?utm_source =feedburner&utm_medium=feed&utm_campaign=Feed%3A+schaechter+%28Small +Things+Considered%29; Maribeth O. Hassett, Mark W. F. Fischer, and Nicholas P. Money, "Mushrooms as Rainmakers: How Spores Act as Nuclei for Raindrops," *PLoS ONE* 10, no. 10 (2015): e0140407, doi.org/10.1371/journal.pone.0140407; Jing Sun et al., "The Abundant Marine Bacterium *Pelagibacter* Simultaneously Catabolizes Dimethyl-sulfoniopropionate to the Gases Dimethyl Sulfide and Methanethiol," *Nature*

Microbiology 1, no. 8 (2016): 16065, doi.org/10.1038/nmicrobiol.2016.65; and Andrew R. J. Curson et al., "Dimethylsulfoniopropionate Biosynthesis in Marine Bacteria and Identification of the Key Gene in This Process," *Nature Microbiology* 2 (2017): 17009, doi.org/10.1038/nmicrobiol.2017.9. Lynn Margulis and Dorion Sagan's quote is from their book *Garden of Microbial Delights.*

John Ingraham wrote that biology creates Earth's chemistry in *March of the Microbes.* To understand the notion of a self-regulating planet, I read James Lovelock, *Gaia: A New Look at Life on Earth* (Oxford: Oxford University Press, 2000) and watched "Daisyworld Pt. 3: Self Regulation," YouTube video, 7:28, posted by e94band, youtube .com/watch?v=1gIQShSrk1I. I also used information from James Lovelock, "Gaia: The Living Earth," *Nature* 426 (2003): 769–70; "2001 Amsterdam Declaration on Earth System Science," IGBP, accessed November 5, 2017, igbp.net/about/history/2001amsterda mdeclarationonearthsystemscience.4.1b8ae20512db692f2a680001312.html; and two books, *Darwin's Blind Spot* by Frank Ryan, and Maureen A. O'Malley, *Philosophy of Microbiology* (Cambridge, England: Cambridge University Press, 2014). My thinking about awe as arising from a sense of oneself on the universal scale was influenced by an interview with Mary Evelyn Tucker of Yale University; her documentary film, *Journey of the Universe,* a collaboration with Brian Thomas Swimme, professor at the California Institute of Integral Studies in San Francisco, journeyoftheuniverse.com; the book by John Grim and Mary Evelyn Tucker, *Ecology and Religion* (Washington, DC: Island Press, 2014); and the geoecologist Stan Rowe's essay, "An Earth-Based Ethic for Humanity," *Natur und Kultur: Transdisziplinäre Zeitschrift für ökologische Nachhaltigkeit* 1, no. 2 (2000): 106–120, ecospherics.net/pages/RoweEarthEthics.html.

Moselio Schaechter asked his students to defend the statement that all living things are connected to other living things, and he blogged about it: "How All Living Things Are Connected (In an Exam)," *Small Things Considered* (blog), 2016, schaechter .asmblog.org/schaechter/2016/05/how-all-living-things-are-connected-in-an-exam.html? utm_source=feedburner&utm_medium=feed&utm_campaign=Feed%3A+schaechter +%28Small+Things+Considered%29.

Would that we all could learn from Moselio.

Index

Gaia hypothesis, 218–19
Gastritis, 173–74
Genes
    Alzheimer's disease and, 195
    of bacteria, 21–22, 106
    diet, effect of, 154
    disease heredity and, 176–77
    disease resistance and, 104, 132–33
    expression of, 41, 58, 75
    of gut bacteria, 153–54
    inherited traits and, 104–5, 115
    junk, 58
    mapping, 103
    in mitochondria, 57
    of plants, modification of, 107, 132–33
    probability and, 104–5
    sexual vs. asexual breeding and, 75–76
Genetically modified organisms (GMO), 107,
    132–33
Genetic transference, 21–22
Genomic species, 42–43
Germophobia, 182–84
Germ theory, 38, 182
Global warming and climate change, xv, 124–
    25, 136, 157–58
Glomalin, 92, 112
Glyphosate, 113, 133–34
GMOs, 107, 132–33
Grain consumption, 117–18, 153, 194
Gulf of Mexico, dead zone in, 112
Gut microbiome
    acquisition of, 127, 158, 162–66, 185
    analysis of, 168–69
    antibiotics, effect of, 159, 166, 178
    changing, 166
    children and, 162–67, 177
    diet for, 191–94
    digestion and, 145–46, 173–74
    diversity of, 169–70, 193–94
    as ecosystem, 144–45
    environmental effects on, 127–28
    foods promoting health of, 189–93
    glyphosate, effect on, 133
    imbalances in, 146
    immune system and, 174–76
    inflammation and, 218
    international comparisons of, 169–70
    neurochemical production and, 72, 177–78
    nutritional content of food and, 145
    obesity and, 192–94
    probiotics for, 188–90

Haber–Bosch process, 111
Handwashing, 183–84
Herbicides, 113, 133–34
Hookworm, 95

Hormones
    for appetite control, 174
    gut microbes, effect on, 72
    of plants, 128
    skin microbes and, 185
    stress and, 72, 177
Household microbiomes, 173, 185, 206–10
Human Genome Project, 103
Human microbiomes. See also Gut microbiome;
        Skin microbiome
    acquisition of, 127–28, 158, 162–66
    microbial clouds, 206
    penis microbiome, 163
    sharing, 185, 205–7
    vagina microbiome, 162–63
Humus, 101
Hygiene hypothesis, 130–31
Hyperthermophiles, 27

Illness. See Disease
Immune system
    autoimmune diseases and, 176
    disease susceptibility and, 196
    of eyes, 165
    gut microbes and, 174–76
    hookworms and, 95
    hygiene hypothesis and, 130–31
    inflammation regulation and, 176–77, 218
    microbiome and, 127, 160
    of newborns, 164
    pets' effect on, 208
    of plants, 130–31
Infection. See Antibiotics; Disease; specific
        types
Inflammation
    acne as, 186
    colon health and, 192
    of gums, 173
    immune system regulation and, 176–77, 218
    skin conditions and, 184
Insects
    decomposition and, 101
    as disease vectors, 39, 70
    pesticides for, 116
    skin microbiome and, 184
    as soil shredders, 95
Iron, 29–30

Lactobacillus, 163, 164–65, 187–88, 194
Large intestine, microbes in, 174
Leprosy, 195–96
Lichens, 87–88
Life. See also Evolution
    carbon and, 12–13
    categorization of, 34–37, 39–44
    chemical basis of, 8